RESIDENTIAL HEATING OPERATIONS AND TROUBLESHOOTING

John E. Traister

Prentice-Hall, Inc., Englewood Cliffs, New Jersey 07632

Library of Congress Cataloging in Publication Data

Traister, John E. (date)
Residential heating operations and troubleshooting.

Includes index.
1. Dwellings—Heating and ventilation. I. Title.
TH7222.T73 1985 697 84-26357
ISBN 0-13-774696-2

Editorial/production supervision
and interior design: *Lisa Schulz*
Jacket design: *Ben Santora*
Cover design: *Edsal Enterprises*

Printed in the United States of America

10 9 8 7 6 5 4 3 2 1

ISBN 0-13-774696-2 01

PRENTICE-HALL INTERNATIONAL, INC., *London*
PRENTICE-HALL OF AUSTRIALIA PTY. LIMITED, *Sydney*
EDITORA PRENTICE-HALL DO BRASIL, LTDA., *Rio de Janeiro*
PRENTICE-HALL CANADA INC., *Toronto*
PRENTICE-HALL HISPANOAMERICANA, S.A., *Mexico*
PRENTICE-HALL OF INDIA PRIVATE LIMITED, *New Delhi*
PRENTICE-HALL OF JAPAN, INC., *Tokyo*
PRENTICE-HALL OF SOUTHEAST ASIA PTE. LTD., *Singapore*
WHITEHALL BOOKS LIMITED, *Wellington, New Zealand*

CONTENTS

PREFACE

With the exception of very large custom-built homes, apartment buildings, and tract-development houses, the average residential heating system of the past has not been complex enough to justify the expense of preparing complete working drawings and specifications. Usually such systems were laid out by the architect in the form of notes on the drawings—after calculating heat loss and gain—or else were laid out by the mechanical contractor on the job, often only as the work progressed. However, many technical developments in residential heating, cooling, and ventilating systems for residential use have greatly expanded the demand and complexity of today's residential HVAC systems. Such relatively new systems as high-velocity duct systems, solar supplemental heat, and the like demand a better and more precise design scheme prior to starting the construction than ever before.

Architects, consulting engineers, mechanical contractors, and their employees are finding a greater and more specialized engineering knowledge for the modern comfort-conditioned residence to be required. Homeowners and repairmen are also finding that a greater knowledge of heating terms and characteristics are needed to properly service and maintain modern residential heating systems. Traditionally, such knowledge could only be obtained by studying such topics as thermal dynamics, hydronics, and the like, requiring many months of theory before any practical applications could be put to use.

What would be useful is a book that completely describes the various heating systems available, with a minimal amount of theory, and quickly gets into practical applications of installation, maintenance, troubleshooting, and repair. *Residential Heating Operations and Troubleshooting* is designed to do just that.

This book will help the mechanical engineering student supplement the theories that are normally taught during his/her first few semesters. It will be very helpful to all building construction personnel, regardless of their trade or profession. Do-it-yourself homeowners will want a copy close by at all times to enable them to take care of the many problems that will eventually arise in any residential heating system.

A deep and grateful bow should be made in the direction of those who helped with the preparation of the manuscript for this book, including the many manufacturers who supplied much of the reference material. A special thanks goes to Ruby Updike, who did much of the research for and preparation of the final manuscript.

John E. Traister

1

INTRODUCTION TO RESIDENTIAL HEATING

Few modern conveniences offer the comfort that one gets from automatic heating and cooling systems in the home. In fact, it is possible to set a desired condition on one's control or thermostat and practically forget about the system. If surrounding conditions are below the desired temperature, the system calls for heat and produces just the right amount. Conversely, if the home is too hot, the system switches automatically to the cooling cycle. Furthermore, the air is cleaned electronically and a prescribed amount of fresh air is introduced to the system.

The energy source for heat was mainly wood up until the past 100 years or so. Now many types of systems are in use and utilize practically every type of fuel conceivable, including the following:

1. Wood
2. Coal
3. Oil
4. Gas
5. Electric
6. Solar
7. Combinations of the above

Of course, other heat sources are also used to a limited degree in industry, but the ones listed above are those used most commonly in residential heating and cooling systems.

TYPES OF RESIDENTIAL CONSTRUCTION

People involved in the design of heating, ventilation, and air-conditioning (HVAC) systems for residential building construction must be able to visualize the building structure and its relation to the HVAC systems in order to plan and coordinate the layout of the equipment, ductwork, and the like. Therefore, HVAC or mechanical designers should have, or acquire, a thorough knowledge of building construction for all types of building structures and should be able to interpret the drawing or plans in terms of the completed project with all of its necessary components.

In general, the following are the basic types of residential building construction:

1. Wood frame
2. Masonry
3. Reinforced concrete
4. Prefabricated structures

Usually, two or more basic types of construction are incorporated into one building. For example, the basement foundation may consist of reinforced concrete while the upper portion of the building is wood frame with masonry veneer.

WOOD-FRAME STRUCTURES

The most common form of residential building construction is the wood-frame type shown in Fig. 1-1. However, the finish on the outside walls of wood-frame buildings consists of many different materials, including masonry, fiberboard, plywood, and several other materials.

In general, wood framing consists of jointing vertical 2 x 4 members to plates and headers—the entire structure then rests on a solid foundation. The interior walls of the structure are usually finished with plaster, drywall, wood panels, or a number of other finishes. The outside walls are first insulated with some type of fiberboard and then finished with wood or masonry. The void spaces between the interior and exterior walls are filled with some type of insulation.

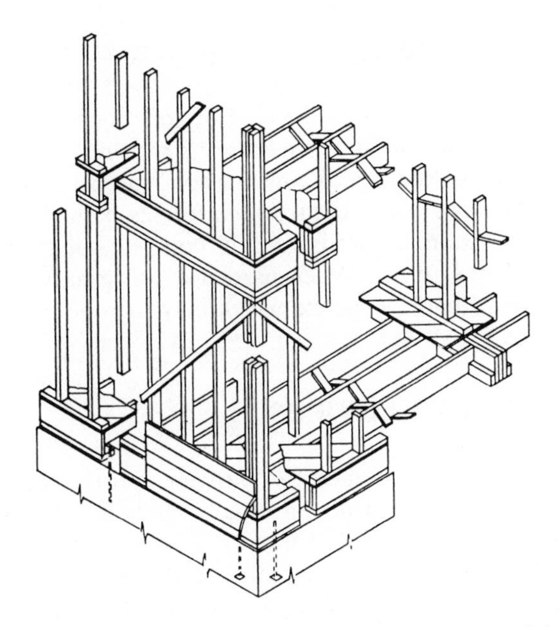

Figure 1–1. Example of a wood-frame building. (*Courtesy of the author*)

MASONRY STRUCTURES

The masonry structure is constructed by placing clay bricks, stones, cement blocks, or other materials, one upon the other, and bonding them together with cement mortar. Except for basement floor slabs, the floor, the ceiling, and the roof construction usually have a wood frame (see Fig. 1–2).

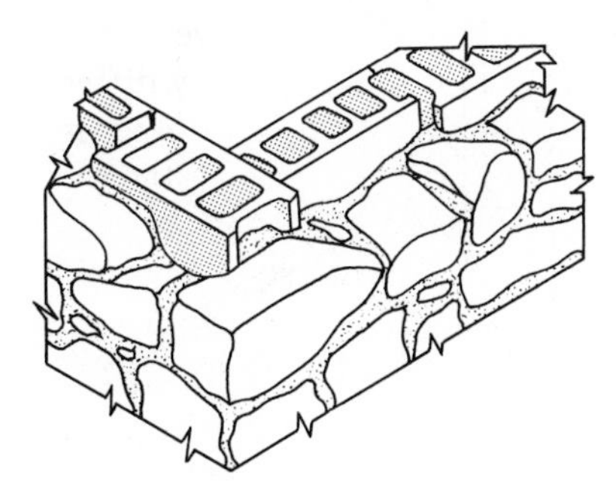

Figure 1–2. Typical masonry construction used for a residential foundation wall. (*Courtesy of the author*)

REINFORCED CONCRETE

Reinforced-concrete construction is the type of construction sometimes used for residential foundations, although masonry structures of concrete blocks are more common today. Reinforced-concrete construction requires the building of forms, in which are placed the steel reinforcing bars or mesh necessary to reinforce the foundation, the walls, and the concrete slab. Concrete is then poured into or onto the forms. When hardened sufficiently, the forms are stripped off to reveal "bare" concrete which may then be painted or left in its natural state (see Fig. 1-3).

PREFABRICATED STRUCTURES

Prefabricated structures have been used extensively over the past few years for residential buildings. Such construction usually has a wood frame with plywood exterior sheathing and drywall interior sheathing. Sections of floors, walls, and roofs are constructed at a central construction factory and then shipped to the building site, where they are assembled by building-trade workers. In some instances, as in the case of module homes, the ductwork and related heating and cooling equipment are also installed in the factory.

RESIDENTIAL HVAC SYSTEMS

All types of HVAC systems used in residential construction will be described in detail in the chapters to follow. However, at this point, the reader should have a general understanding of each. Therefore, a brief description follows of the ones most commonly used.

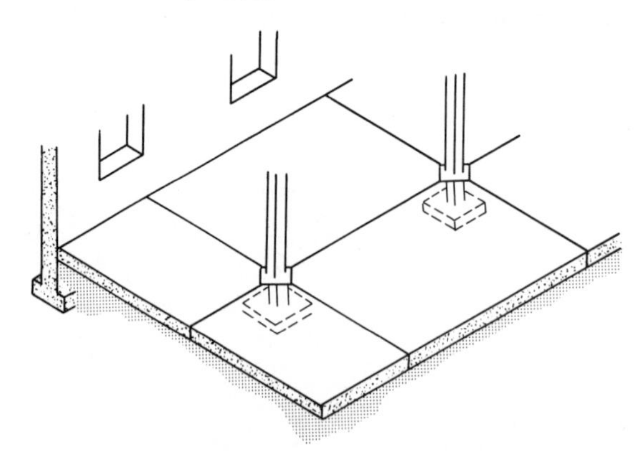

Figure 1-3. Example of reinforced concrete used for basement wall and basement floor. (*Courtesy of the author*)

Hot-Water Baseboard Heating: A zone hydronic (hot-water) heating system permits selection of different temperatures in each zone of the home. Baseboard heaters located along the outer walls of the rooms provide a blanket of warmth from floor to ceiling. Hot water is supplied from a central heating unit, which also supplies domestic hot water through a separate circuit. Hot-water boilers for the home are normally manufactured for use with oil, gas, or electricity for the fuel source.

The chief disadvantage of hot-water systems is that they do not use ductwork, but piping. Therefore, if a central air-conditioning system is also desired, separate ductwork must be provided.

Electric Baseboard Heaters: Baseboard units utilize resistant heat as their basis of operation and have several advantages over other types of heat:

1. Fuel tanks and chimneys are not needed.
2. The initial installation cost is low.
3. Each room may be controlled separately by its own thermostat.

However, there are also some disadvantages with this type of system:

1. Humidification is hard to control.
2. Heating units require constant cleaning to eliminate dust, which will burn and stain walls above heaters.

Combination Heating/Cooling Units: The advantages of through-wall heating and cooling units are similar to those of electric baseboard heaters except that they also have cooling capability.

Heat Pumps: A heat pump is a system in which refrigeration equipment takes heat from a source and transfers it to a conditioned space (when heating is desired), and removes heat from the space when cooling and dehumidification are desired.

Central Heating Units: Central heating and cooling units distribute heating and cooling from a centrally located source, by means of circulating air. Heating fuels used include oil, gas, coal, wood, and electricity. The air is distributed via air ducts run throughout the building.

Infrared Heaters: These are assemblies that make use of the heat output of infrared lamps or other sources. Such heaters provide fast-response, high-temperature radiation and are particularly suited for use in locations in which it is difficult or impractical to maintain air temperatures at comfortable levels.

Fan-Driven Forced-Air Unit Heaters: The heat source may be provided by either electric resistance heating or piped steam or hot water. Cooling may also be used in combination with the heat when cold water from a chiller is circulated through the coils of the unit.

High-Velocity Heating and Cooling Systems: High-velocity systems have been used in commercial buildings for quite some time, but due to the noise once common to this type of system, very few were installed in residential buildings. Now, however, new designs in this type of system have reduced the noise level to the point where it is quite acceptable for residential applications.

The compactness of the equipment and the small ductwork and outlets make this system well suited for use in existing structures where the installation of conventional forced-air ducts would require too much cutting and patching. On the other hand, the small 3½-in.-diameter ducts of a residential high-velocity system can be "fished" through wall partitions, corners of closets, and similar places. The system's small air outlets (only 2-in. openings) can be placed nearly anywhere and still provide good air distribution.

Solar Heat: A basic solar system consists of a solar (heat) collector that is usually arranged so that it faces south, a heat reflector mounted on the ground in front of the collector, a storage tank to hold the sun-heated water, a circulating pump, and piping. The operation of these components is simple. The storage tank is filled with water which is pumped to the top of the heat collector. As the water flows over the collector, the sun heats it. At the bottom of the collector, the heated water is collected in a trough and then flows back to the storage tank.

The water may be pumped from the storage tank through a system of pipes to baseboard radiators in the living area; a heat exchanger in the water loop may transfer the heat to forced air for distribution via a duct system; or a heat pump may perform the exchange, using the water loop as both heat source and sink.

Radiant Heating Cable: Embedded heating cable is the most common large-area electric ceiling heating system in use today. The heating cable is laid out on a grid pattern and stapled to a layer of gypsum lath on the ceiling joists. Then the cable is covered with a layer of wet plaster or another layer of gypsum board is placed over the cable. Heat cable is available on reels for easy payout as the grid is formed. The nonheating leads on the ends of the run of heating cable are brought up through the plates and fed down through the wall to the control thermostat or relay. Cables may be obtained in many lengths and spaced to satisfy room heating capacity requirements.

Miscellaneous Heaters: Since about 1973 much experimenting has been done with alternative heating sources and of these, the wood furnace seems to have gained the most popularity in certain areas. Modern wood furnaces are automatically controlled and most require filling only once each day. Their efficiency is high and where a plentiful wood supply is available at reasonable cost—such as for rural residents with a woodlot—these systems are difficult to surpass when it comes to saving energy and heating expenses. Although usually more expensive to install than conventional models, wood furnaces will quickly pay for themselves in fuel savings if, as mentioned previously, a cheap source of wood is available.

These and other types of heating, cooling, and ventilating systems will be discussed in more detail in the chapters to follow, with particular emphasis on their design and application.

COMBUSTION

In dealing with residential heating systems, it is useful to have an elementary knowledge of combustion and fuels. In general, *combustion* is the very rapid chemical combination of two or more elements, accompanied by the production of light and heat. The atoms of some of the elements have a great attraction for those of other elements, and when they combine they rush together with such rapidity and force that heat and light are produced. For example, oxygen, which has a great attraction for nearly all the other elements, has a particular fondness for carbon, and whenever these two elements come into contact at a sufficiently high temperature, they combine with great rapidity. The combustion of coal in a furnace is of this nature. The temperature of the furnace is raised by kindling the fire, and then the carbon of the coal begins to combine with oxygen taken from the air.

When carbon and oxygen combine they form carbon dioxide, CO_2; when hydrogen and oxygen combine they form water, H_2O. These are called the *products of combustion.* The oxygen required for combustion is usually obtained from the air, which is a mixture composed of approximately 23 parts of oxygen and 77 parts of nitrogen by weight. The nitrogen that enters the furnace with the oxygen takes no part in the combustion, but passes through the furnace and up the chimney without any change in its nature.

Air Required for Combustion: When carbon is burned to carbon dioxide, CO_2, 1 atom of carbon unites with 2 atoms of oxygen. Carbon has an atomic weight of 12 and oxygen has an atomic weight of 16, so that the molecular weight of CO_2 is $(1 \times 12) + (2 \times 6) = 44$; CO_2 is thus composed of $12/44 = 27.27\%$ carbon and $32/44 = 72.73\%$ oxygen. To burn a pound of carbon to CO_2, therefore, requires $32/12 = 2\frac{2}{3}$ lb of oxygen. If the oxygen is taken from the air, it will take $2\frac{2}{3}/0.23 = 11.6$ lb of air to supply the $2\frac{2}{3}$

lb of oxygen. This is because only 23% of air is oxygen. Therefore, 1 lb of carbon requires 11.6 lb of air for complete combustion. Of this air, 2.67 lb is oxygen, which combines with the pound of carbon to form 3.67 lb of carbon dioxide. The 8.93 lb of nitrogen contained in the air passes off with the CO_2 as a product of combustion.

In dealing with the complete combustion of 1 lb of hydrogen, the product of the combustion is water, H_2O. H_2O is composed by weight of 2 parts hydrogen to 16 parts oxygen, requiring $16/2 = 8$ lb of oxygen to unite with it. The air required to furnish 8 lb of O is $8/0.23 = 34.8$ lb.

Incomplete Combustion: There is one other case that may occur: the combustion of carbon may not be complete. If insufficient air or oxygen is supplied to the burning carbon, it is possible for the carbon and oxygen to form another gas, carbon monoxide, CO, instead of carbon dioxide, CO_2. The combustion of 1 lb of carbon to form CO, of course, requires only one-half the oxygen that would be necessary to form CO_2. This is because in CO gas 1 atom of carbon seizes 1 atom of oxygen instead of 2. To burn 1 lb of carbon to CO_2 requires 11.6 lb of air. To burn it to CO would therefore require but 5.8 lb of air.

Calorific Value of Fuels: The amount of heat, in British thermal units (Btu), developed by the complete combustion of 1 lb of a fuel is called the *calorific value* of that fuel; it is also sometimes known as the *heat value* or the *heat of combustion*. It may be determined most accurately by burning a known weight of the fuel with oxygen in an instrument known as a *calorimeter*. The gases resulting from the combustion are passed through a known weight of water and give up their heat to the water. By noting the rise of temperature of the water, it is possible to calculate the amount of heat absorbed, and thus to determine the heat that would be produced by the combustion of 1 lb of the fuel. The calorific values of the elements most commonly found in fuels are as follows:

	Btu/lb
Hydrogen, burned to water, H_2O	62,000
Carbon, burned to CO_2	14,600
Carbon, burned to CO	4,400
Sulfur, burned to SO_2	4,000

If the various percentages, by weight of the elements, composing a fuel are known, the approximate calorific value of that fuel may easily be calculated by the formula

$$X = 14{,}600C + 62{,}000\,(H - O/8) + 4000S$$

where X = calorific value of fuel, in Btu per pound
C = percentage of carbon, expressed as a decimal
H = percentage of hydrogen, expressed as a decimal
O = percentage of oxygen, expressed as a decimal
S = percentage of sulfur, expressed as a decimal

For example, if a coal contains 85% carbon, 4% oxygen, 6% hydrogen, 1% sulfur, and 4% ash, the heat of combustion per pound may be found by the following formula:

$$X = 14{,}600 \times 0.85 + 62{,}000\,(0.06 - 0.04/8) + 4000 \times 0.01 = 15{,}860 \text{ Btu}$$

FUELS

Fuels for Heat: Many fuels are used to produce heating for residential use, but those in most common use are coal, wood, petroleum, and natural and propane gas. In some parts of the world, other fuels, such as waste gases from blast furnaces, straw, bagasse, dried tan bark, green slabs, sawdust, and peat, are also used. All these fuels are composed either of carbon alone or carbon in combination with hydrogen, oxygen, sulfur, and noncombustible substances.

Classes of Coal: The different varieties of coal may be classed in four main groups: anthracite, semianthracite, semibituminous, and bituminous.

Anthracite coal contains from 92.31 to 100% fixed carbon and from 0 to 7.69% volatile hydrocarbons. It is rather difficult to ignite and a strong draft is required to burn it. It is quite hard and shiny; in color it is a grayish black. It burns with almost no smoke, and this fact gives it a peculiar value in places where smoke is objectionable. Anthracite coal is known to the trade by different names according to the size into which the lumps are broken. These names, with the generally accepted dimensions of the screens over and through which the lumps of coal will pass, are as follows. Culm passes through 3⁄16-in. round mesh. Rice passes over 3⁄16-in. mesh and through 1⁄4-in. square mesh. Buckwheat No. 2 passes over 1⁄4-in. mesh and through 5⁄16-in. mesh. Buckwheat No. 1 passes over 5⁄16-in. mesh and through 1⁄2-in. square mesh. Pea passes over 1⁄2-in. mesh and through 3⁄4-in. square mesh. Chestnut passes over 3⁄4-in. mesh and through 1 3⁄8-in. square mesh. Stove passes over 1 3⁄8-in. mesh and through 2-in. square mesh. Egg passes over 2-in. mesh and through 2 3⁄4-in. square mesh. Broken passes over 2 3⁄4-in. mesh and through 3 1⁄2-in. square mesh. Steamboat passes over 3 1⁄2-in. mesh and out of screen. Lump passes over bars set from 3 1⁄2 to 5 in. apart.

Semianthracite coal contains from 87.5 to 92.31% fixed carbon and from 7.69 to 12.5% volatile hydrocarbons. It kindles easily and burns more freely than true anthracite coal, hence is highly esteemed as a fuel. It crumbles readily and may be distinguished from anthracite coal by the fact that when just fractured it will soil the hand, whereas anthracite will not do so. It burns with very little smoke. Semianthracite coal is broken into various sizes for the market; these sizes are the same and are known by the same names as those of the corresponding sizes of anthracite coal.

Semibituminous coal contains from 75 to 87.5% fixed carbon and from 12.5 to 25% volatile hydrocarbons. It differs from semianthracite coal only in having a smaller percentage of fixed carbon and more volatile hydrocarbons. Semibituminous and bituminous coals are known to the trade by the following names. Lump coal includes all coal passing over screen bars 1½ in. apart. Nut coal passes over bars ¾ in. apart and through bars 1½ in. apart. Pea coal passes over bars ⅜ in. apart and through bars ⅜ in. apart. Slack includes all coal passing through bars ⅜ in. apart.

Bituminous coal contains from 0 to 75% fixed carbon and from 25 to 100% volatile hydrocarbons. It may be divided into three classes, whose names and characteristics are as follows. *Caking coal* is the name given to coals that when burned in the furnace, swell and fuse together, forming a spongy mass that may cover the entire surface of the grate. These coals are difficult to burn, because the fusing prevents the air from passing freely through the bed of burning fuel. When caking coals are burned, the spongy mass must be broken up frequently with a slice bar to admit the air needed for its combustion. *Free-burning coal* is a class of bituminous coal that is often called *noncaking coal* from the fact that it has no tendency to fuse together when burned in a furnace. *Cannel coal* is a grade of bituminous coal that is very rich in hydrocarbons. The large percentage of volatile matter makes it valuable for gas making.

Lignite, or *brown coal,* contains from 30 to 60% carbon, a small quantity of hydrocarbons, and a large amount of oxygen. It occupies a position between peat and bituminous coal, being probably of later origin than the latter. It has an uneven fracture and a dull luster. Exposure to the weather causes it to absorb moisture rapidly, and it will then crumble quite readily. It is noncaking and yields only moderate heat, and is in this respect inferior to even the poorer grades of bituminous coal.

Miscellaneous Fuels: *Coke* is made from bituminous coal by driving off the volatile matter. It consists of from 88 to 95% carbon, ¼ to 2% sulfur, and from 4 to 12% ash. It is seldom used for residential use.

Wood, of course, is used for fuel in localities where it is plentiful. It contains from 20 to 50% moisture when cut, and this percentage is not reduced much below 20% by drying. Wood has a calorific value of 6000 to 7000 Btu/lb.

Peat consists of vegetable matter that is partly carbonized and is found at the surface of the earth. It contains from 75 to 80% water when cut, and must be dried before it can be used as fuel.

Bagasse is the refuse left after the juice has been extracted from sugarcane by means of rolls. It is used to some extent in tropical and semitropical countries. Its use is limited to places where sugarcane is grown.

The use of dried tan bark, straw, slabs, and sawdust is local in nature and usually confined to tanneries, planing and sawmills, and threshing outfits.

It has been stated that on the average 1 lb of good bituminous coal may be considered as the equivalent of 2 lb of dry peat, 2½ lb of dry wood, 2½ to 3 lb of dry tan bark or sun-dried bagasse, 3 lb of cotton stalks, 3¾ lb of straw, 6 lb of wet bagasse, and from 6 to 8 lb of wet tan bark.

Liquid Fuel: The fuel employed most extensively in the generation of electricity is coal, the most valuable of the solid fuels. In some parts of the world, however, it has been found convenient and economical to use liquid fuel. This is obtained chiefly from petroleum, which is a natural oil obtained from the earth. In its original state it is usually dark green in color when viewed in the sunlight, but when held up to the light so that the light passes through it, it is reddish brown. The appearance of oil will vary somewhat depending on the locality from which it is derived. In some cases it is almost as clear and colorless as water, and in other cases it is black, but American petroleum is commonly brown or reddish brown with a green luster.

Petroleum in the form in which it issues from the earth is known as *crude oil.* It usually contains 83 to 87% carbon, 10 to 16% hydrogen, and small percentages of oxygen, nitrogen, and sulfur. Some crude oils are devoid of sulfur and nitrogen, but all those obtained along the Pacific coast contain oxygen, sulfur, nitrogen, and a small percentage of moisture. The presence of sulfur in an oil is manifested by a very disagreeable odor. The following analyses of crude oils from Texas and California indicate the composition of the oils from these fields.

Constituents of Crude Oil	Texas %	California %
Carbon	84.60	85.0
Hydrogen	10.90	12.0
Sulfur	1.63	0.8
Oxygen	2.87	1.0
Nitrogen	—	0.2
Moisture	—	1.0

Owing to the great demand for gasoline in all its grades, the better grades of petroleum are treated to recover the lighter hydrocarbons. Of these, gasoline is among the early distillates, and when the gasoline, naphtha, and kerosene have been separated, the residue contains the lubricating oils, paraffin, and coke. This residue may be further distilled to obtain the several products named, or it may be used as a fuel, then being termed *fuel oil.*

The combustible elements contained in oil fuels are the same as those in coal: carbon and hydrogen and possibly a small proportion of sulfur. The heat of combustion per pound of oil, or calorific value, may be found approximately from the chemical analysis of the oil by the same formula as that used for finding the heat value of coal. A more accurate method, however, is to burn a known weight of oil in a calorimeter and to measure the heat generated, from which the heat per pound of the oil may readily be calculated. From the results of available tests it is found that the heat of combustion per pound of oil fuel varies between 17,000 and 21,000 Btu. The average calorific value of Texas and California crude oils seems to be about 18,600 Btu/lb.

When coal is used, the furnace contains a considerable amount of fuel, but early experiments with liquid fuel soon proved that the methods adopted for solid fuel were not applicable to liquid fuel, and that the latter could not be burned successfully in bulk. To ensure satisfactory burning of oil fuel, it must first be changed to a vapor, and this is accomplished by atomizing the oil, that is, converting it into the form of a very minutely divided spray. It is the vapor that burns, not the liquid oil itself. If a sliver of burning wood is thrust into an open pan of fuel oil, the oil will not ignite, and the flame of the stick will be extinguished. The reason is that insufficient oil surface is exposed to the action of heat, and vaporization does not occur rapidly enough to supply the necessary quantity of inflammable gasses to support combustion. By atomizing the oil, each minute particle is exposed to the air, thus providing for rapid evaporation and complete combustion.

Having changed the oil to a spray or to a vapor, it is next necessary to mix it intimately with air in the correct ratio to produce complete combustion. There are different methods by which the air is admitted so as to accomplish the mixing. Sometimes it is allowed to enter through holes surrounding the spraying devices and sometimes through openings from the ash pit into the furnace; combinations of both methods may also be used. In any event, the main object to be attained is thorough mixing of the spray and the air so that each particle of oil will be surrounded by the oxygen required for its perfect combustion.

2

HOT-WATER HEATING SYSTEMS

Hot-water heating systems are frequently used in residential applications for space heating, that is, to heat the entire home. This type of heating system is considered by many to be the most comfortable system in existence. There are several types of systems available, and although some of those mentioned in the following paragraphs have become somewhat outdated, the reader should be familiar with them to enable repairs to be made. Of course, the more modern systems are also covered.

HOT-WATER HEATING

In a gravity-flow heating system, circulation of hot water to the radiators is accomplished by the difference between the weight of water in the supply main and that in the return main. Figure 2–1 illustrates this principle. Water heated in the boiler increases in volume and rises at the same time that the cooler, heavier water in the system moves downward in the return main, causing a constant circulation to be maintained. A steam-heating system operates in much the same way and such a system is illustrated in Fig. 2–2.

The forced hot-water system employs an electric pump to provide circulation, as shown in Fig. 2–3 (see also Fig. 2–4). In this type of system, circulation is greatly speeded up so that radiators, water coils, and similar units can be supplied with hot water almost instantly whenever it is needed. Also, a constant temperature can be maintained in the system to compensate for outdoor weather conditions.

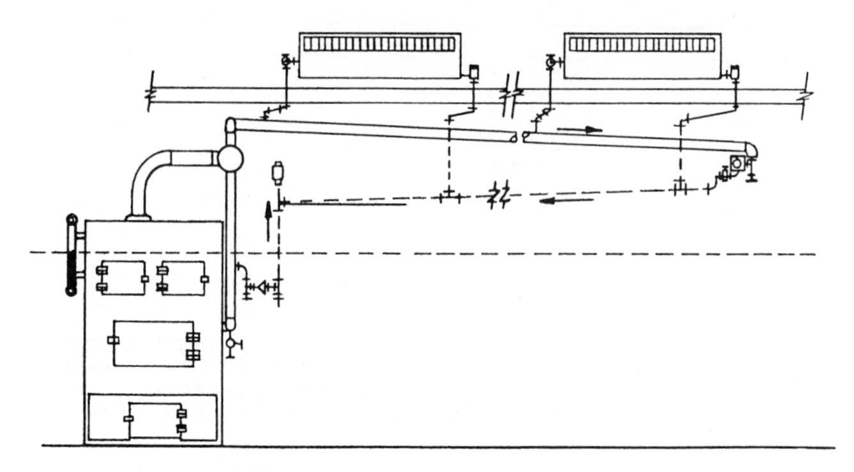

Figure 2–1. Typical gravity-pipe hot-water heating system. Water flow is caused by the difference in weight of the hot water in the supply main and the cooler water in the return main. (*Courtesy of Craftsman Book Company*)

BASIC FACTS ABOUT HEATING

General: A heating system must be designed to replace the heat lost through the roof, walls, and floors of a building and also the heat lost by infiltration of air through cracks around doors and windows. The rate of heat loss is determined by

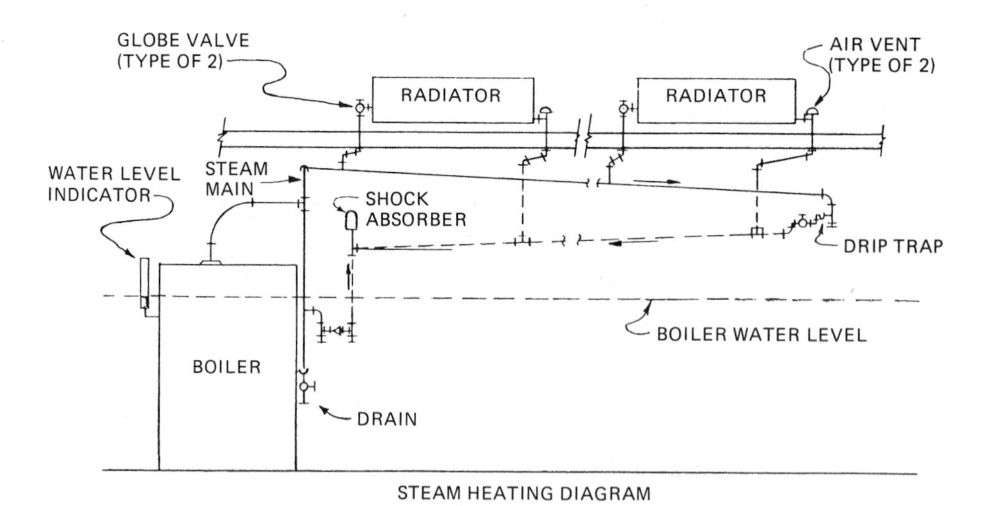

Figure 2–2. System illustrated in Fig. 2–1 as it would appear in a working drawing. (*Courtesy of Craftsman Book Company*)

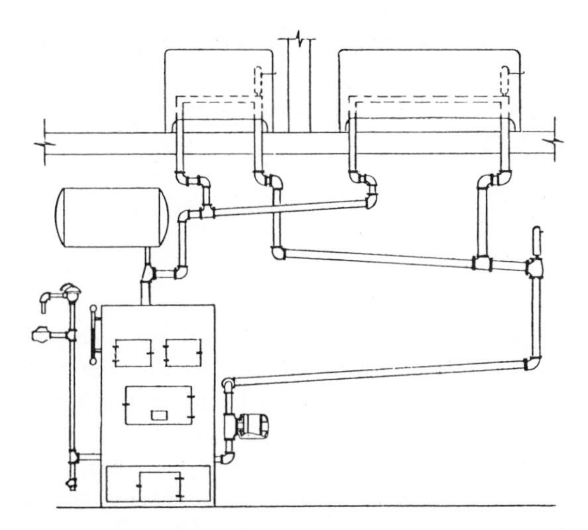

Figure 2–3. Typical two-pipe, forced-circulation, hot-water heating system. *(Courtesy of Craftsman Book Company)*

1. The building construction
2. The velocity of prevailing winds
3. The outdoor temperature

When the proper calculations are performed, the heat loss of the building can be determined and the required capacity of the heating plant can be chosen.

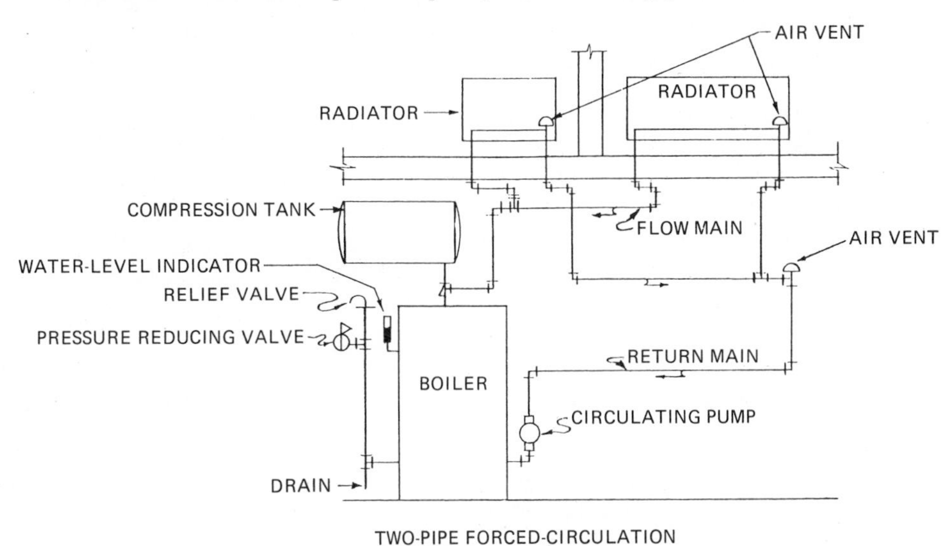

Figure 2–4. System illustrated in Fig. 2–3 as it would appear on a working drawing. *(Courtesy of Craftsman Book Company)*

The accepted measurement of heat is the *British thermal unit* (Btu):

One Btu is the amount of heat required to raise the temperature of 1 pound of water 1 degree Fahrenheit.

Nearly all calculations of heat output and heat input are made by the Btu method.

Pressure Drop: Pressure drop is a term that expresses the fact that power is consumed when liquids are moved through pipes, heating units, fittings, and so forth. Pressure drop is caused by the friction created between the inner walls of the conveyor (pipe, for example) and the moving liquid.

Head Pressure: As used in designing the capacity of a circulating pump, head pressure is the maximum pressure drop against which the pump can induce a flow of liquid. If a certain size of booster pump is connected to a tank of water as illustrated in Fig. 2-5, it will pump water to a height of 7 ft. This height, therefore, is the maximum pressure head against which the pump can cause a flow of water.

One- and Two-Pipe Forced Hot-Water Systems: Forced hot-water systems are classified as one-pipe or two-pipe designs. Two-pipe systems are further divided into direct-return and reverse-return layouts.

In the one-pipe system, illustrated in Fig. 2-6, a single main line in one or more circuits is used to circulate water. Branch pipes or *risers* from this main are equipped with special fittings at their connections to the main line.

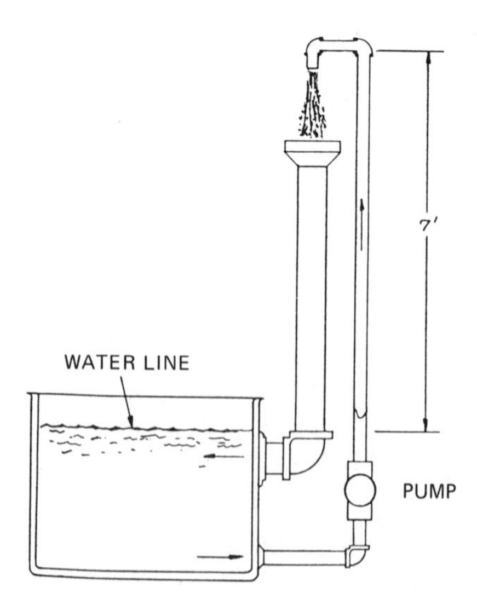

Figure 2–5. Maximum pressure head is the height to which a pump can cause water to flow. *(Courtesy of Craftsman Book Company)*

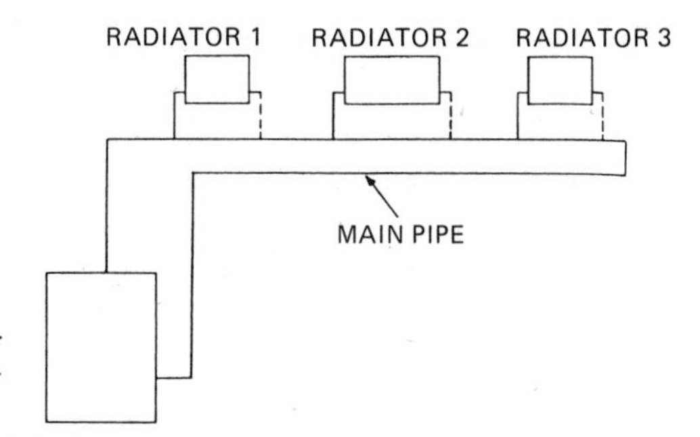

Figure 2–6. One-pipe forced hot-water system. (*Courtesy of Craftsman Book Company*)

These fittings introduce the correct amount of resistance needed to assure proper diversion of hot water into the radiator.

In the reverse-return system of a two-pipe system, illustrated in Fig. 2–7, the first radiator of the main line is the last to return, and all radiator circuits are of equal length. Therefore, the problem of proper radiation balance is greatly simplified in this type of system.

In the direct-return system, illustrated in Fig. 2–8, the first radiator taken off the main line is the first radiator to return its load. The last radiator on the line is the last to return. Consequently, the water circuits to the radiators are of unequal length, and maintaining a proper balance of heat distribution can cause serious difficulties at times.

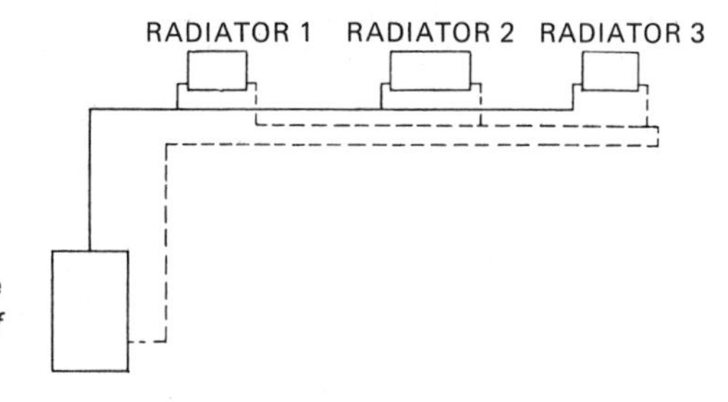

Figure 2–7. Reverse-return two-pipe forced hot-water system. (*Courtesy of Craftsman Book Company*)

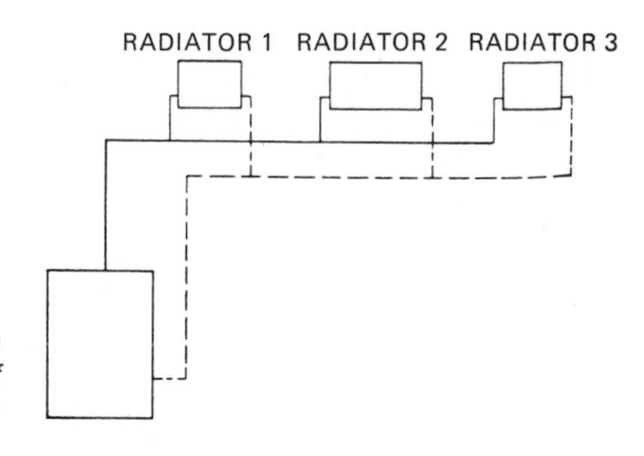

Figure 2–8. Direct-return two-pipe forced hot-water system. (*Courtesy of Craftsman Book Company*)

AIR SUPPLY FOR BOILER ROOMS

All open-flame equipment must be installed in a location in which ventilation facilities permit satisfactory combustion of gas, proper venting, and the maintenance of ambient temperatures at safe limits under normal conditions of use.

The general rules set forth below are based on natural-draft venting and the requirements necessary for proper operation of equipment. They do not take into account the makeup air necessary to supply a building exhaust system. When an exhaust system is installed in a building, provisions should be made for makeup air to prevent negative pressures within the building. These pressures will adversely affect combustion efficiencies of equipment used for heating and water heating.

EQUIPMENT LOCATED IN UNCONFINED SPACES

In unconfined spaces within a building of conventional frame, brick, or stone construction, infiltration will normally provide adequate air to the open-flame equipment. In unconfined space within a building of unusually tight construction, air must be provided from out-of-doors. A permanent opening having a total free area of not less than 1 $in.^2$ per 5000 Btu/h (British thermal units per hour) of total input is required. Ducts that convey this air should have the same cross-sectional area as the free opening, and when rectangular, they should not be less than the cross-sectional area of a 3-in. round duct (7.06 $in.^2$).

EQUIPMENT LOCATED IN CONFINED SPACES

Provisions must be made to supply a means of air circulation when installing equipment in a confined space.

All Air From Inside the Building: The boiler room should be provided with two permanent openings, one near the top of the enclosure and one near the bottom. Each opening should have a free area of not less than 1 $in.^2$ per 1000 Btu/h of total input. These openings are connected with the interior areas of the building that, in turn, must have adequate infiltration from the outside.

All Air From the Outside: The boiler room should have two permanent openings, one in the top and one in the bottom of the enclosure. These openings should communicate directly, or by means of ducts, with the outdoors. When connected directly to the outdoors or when connected by means of vertical ducts, each opening should have a free area of not less than 1 $in.^2$ per 4000 Btu/h of total input. If horizontal ducts are used, each opening should have a free area of not less than 1 in.^{2} per 2000 Btu/h of total input. When rectangular, these ducts should not be less than the cross-sectional area of a 3-in. round duct (7.06 in.^{2}).

Louvers and Grilles: The free areas do not take into consideration the louvers, grilles, or screens in the openings. Openings equipped with wooden louvers will have 20 to 25% free area. Those with metal louvers or grilles will have from 60 to 75% free area. Thus the openings will have to be enlarged to compensate for the blocking effect of louvers and grilles.

The examples of heating systems illustrated in Figs. 2-9 through 2-12 were taken from actual working drawings. Each illustration should be carefully analyzed. The following points are some of the recommended design practices used when planning the layout for residential hot-water heating.

A baseboard radiation system should be installed along the outside walls. The heating elements should be equally distributed under all windows. If the outside walls do not provide sufficient footage for the required radiation, use the inside walls also.

To ensure proper heating satisfaction of a nonzoned single-loop system, the water from the boiler should flow in the following sequence:

1. To the bathroom
2. To the kitchen
3. To the living room
4. To the dining room and bedrooms

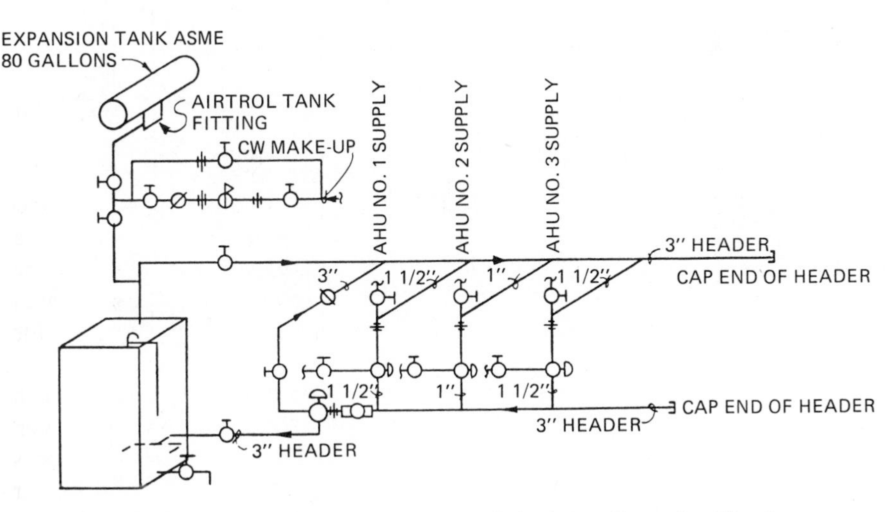

Figure 2–9. Heating system for hot-water coils in air-handling units. (*Courtesy of Craftsman Book Company*)

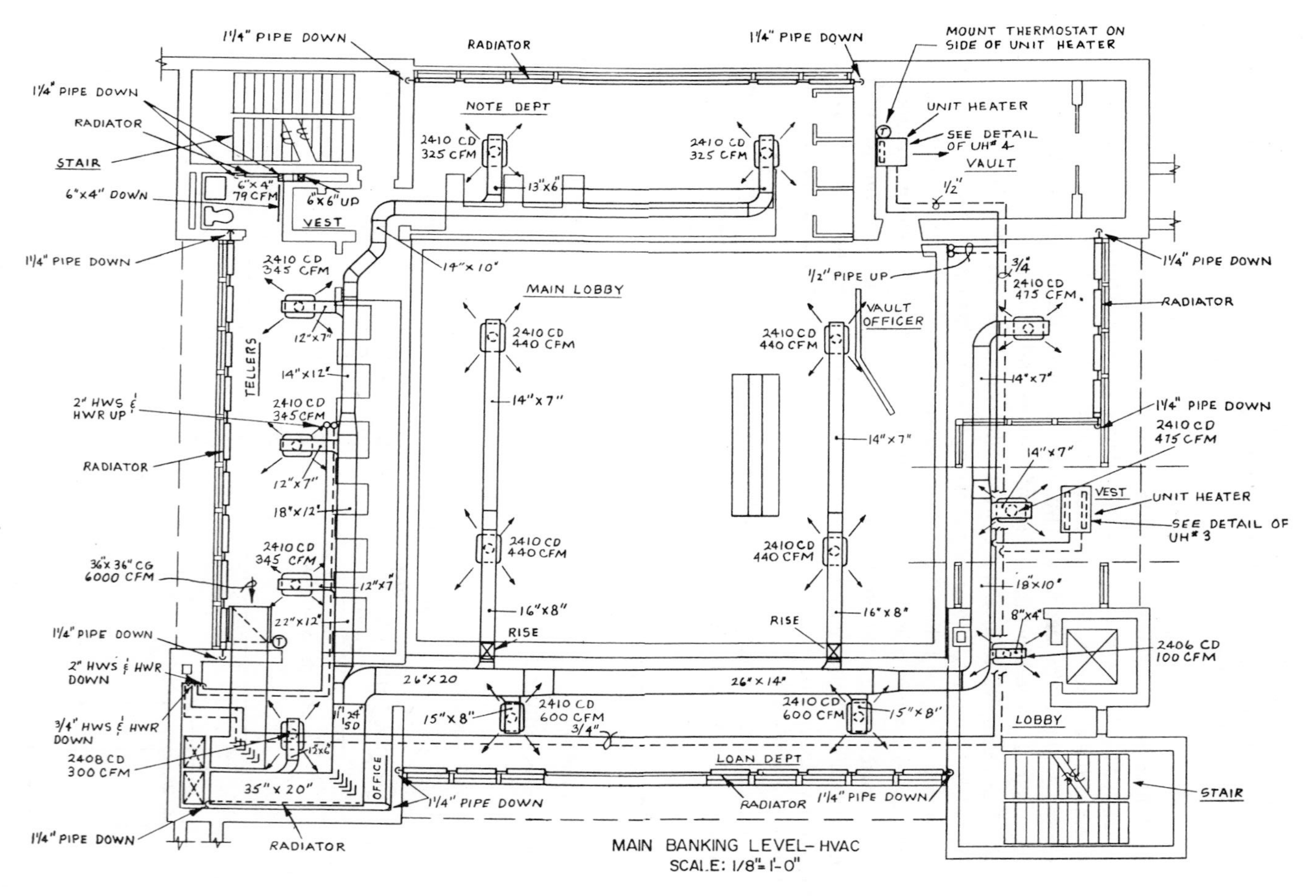

Figure 2–10. Plan view of the HVAC system for the main level of a bank building. (*Courtesy of Craftsman Book Company*)

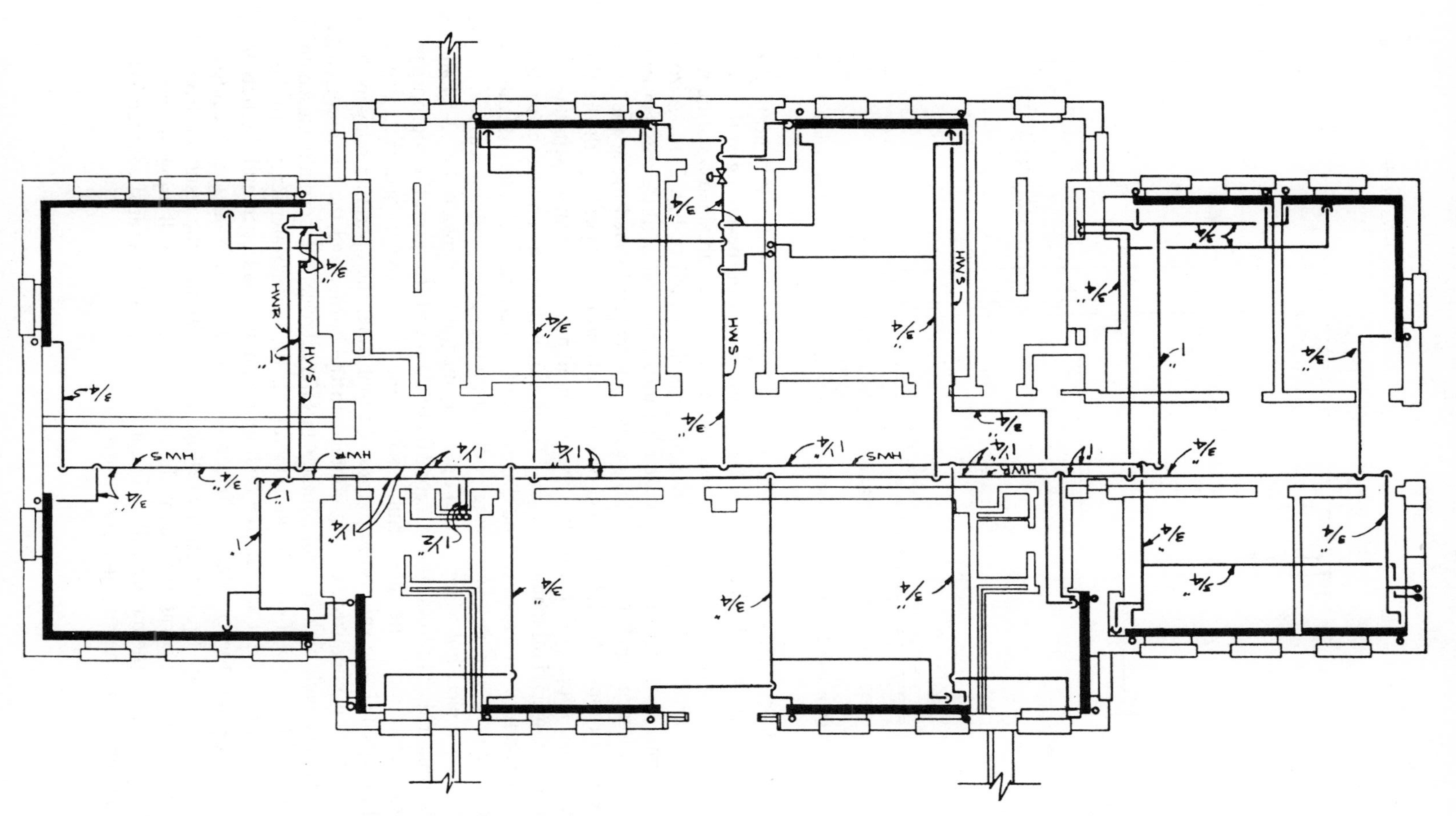

Figure 2–11. Plan view of a heating system for hot-water baseboard heaters in an education building. (*Courtesy of Craftsman Book Company*)

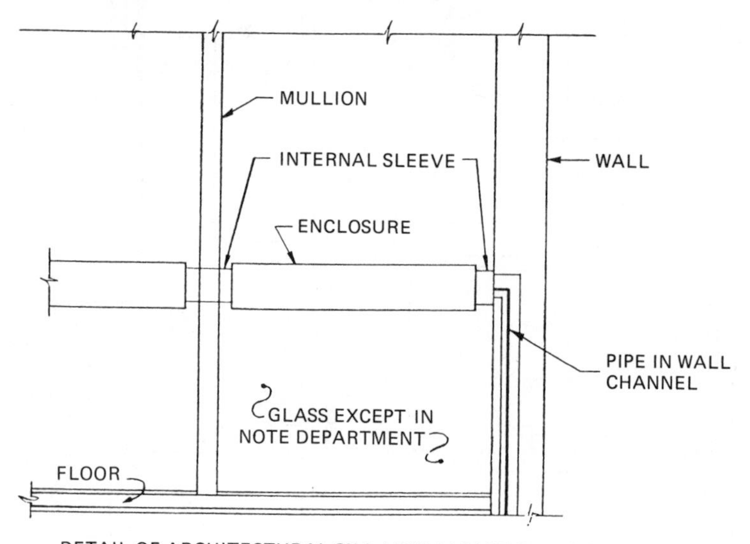

Figure 2–12. Detailed drawing. (*Courtesy of Craftsman Book Company*)

This circuiting procedure will result in the most satisfactory heating installation, due to the fact that the hottest water is at the beginning of the loop. The ultimate in heating satisfaction, however, will be obtained by using motorized valves and "zoning" the system as illustrated in Fig. 2–13. A zone hydronic (hot water) heating system permits selection of different temperatures in each zone of the home. Baseboard heaters located along the outer walls of rooms provide a blanket of warmth from floor to ceiling; the heating unit also supplies domestic hot water simultaneously, through separate circuits. A special attachment coupled to the hot-water unit can be used to melt snow and ice on walkways and driveways in winter, and a similar attachment can be used to heat a swimming pool during the spring and fall seasons.

A typical hot-water system operating diagram is shown in Fig. 2–14 and is explained as follows. When a zone thermostat calls for heat, the appropriate zone valve motor begins to run, opening the valve slowly; when the valve is fully opened, the valve motor stops. At that time, the operating relay in the hydrostat is energized, closing contacts to the burner and the circulator circuits. The high-limit control contacts (a safety device) are normally closed so that the burner will now fire and operate. If the boiler water temperature exceeds the high-limit setting, the high-limit contacts will open and the burner will stop, but the circulator will continue to run as long as the thermostat continues to call for heat. If the call for heat continues, the resultant drop in boiler water temperature—below the high-limit setting—will bring the burner back on. Thus the burner will cycle until the thermostat is satisfied; then both the burner and circulator will shut off.

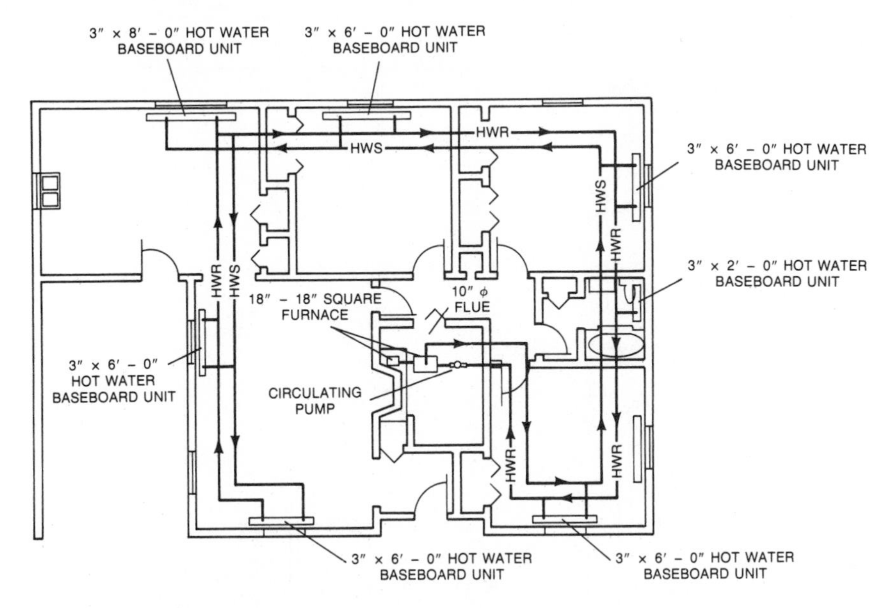

Figure 2–13. Hot-water heating system sketched out on a floor plan. (*Courtesy of Craftsman Book Company*)

Hot-water boilers for the home are normally manufactured for use with oil, gas, or electricity. Although a zoned hot-water system is comparatively costly to install, the cost is still competitive with the better hot-air systems. The chief disadvantage of hot-water systems is that they do not use ducts. If you wish to install central air conditioning, you must install a complete duct system along with the central unit.

HOT-WATER COMBINATION UNITS

Many homes throughout the country utilize older steam or hot-water heating systems which may be in need of repair or replacement. In doing so, the owners may want to consider installing through-wall combination units in place of the radiators. These units are easily installed, and comfort conditioning can then be controlled floor by floor, wing by wing, or room by room. The initial cost of such a system usually is much less than that of a central air-conditioning system offering comparable performance, and the system provides both heating and cooling.

Operating costs for through-wall units are lower than those of many other systems due to the high efficiency of room-by-room control. The living area, for example, can be heated at a temperature of 70°F, while the bedrooms may require only 55°F. Areas that are used only occasionally, such as a guest bedroom or workshop, may be kept at an even lower temperature until occupied.

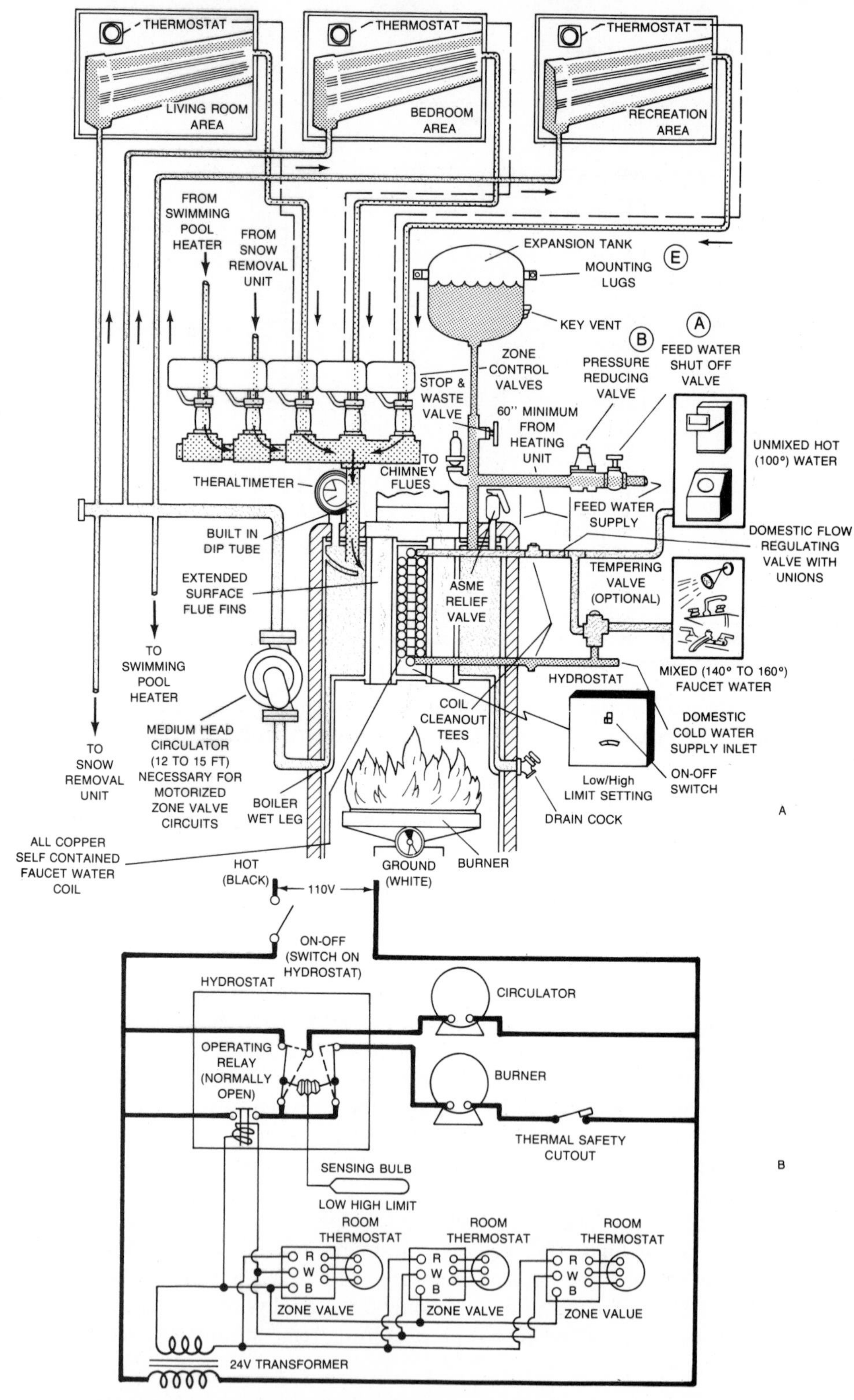

Figure 2–14. (a) Typical hot-water system operating diagram; (b) typical electrical wiring. (*Courtesy of Craftsman Book Company*)

The same principle applies to cooling the areas during warm months. Unlike a central forced-air duct system, through-wall units never need balancing from season to season and if a unit should fail, the defective chassis can be replaced immediately or taken to a shop and repaired.

The main disadvantages of combination heating and cooling units should be considered carefully, however, before making arrangements to go ahead with such a project. First, at each location, you will have a relatively large unit protruding inside your home. There will also be an opening of approximately 15 in. by 54 in. on the outside of your home; this is the grille for air-cooling the air-conditioning compressor and for makeup air. Although these openings are usually flush with the outside of the building, the esthetics of a home may be downgraded in certain cases.

Another consideration is the noise that these units make due to the compressor and blower fan. The better units are very quiet, but some are so noisy that you would not want them in your home. Therefore, you should examine several brands closely before making a decision.

The esthetics of through-wall units can be improved somewhat by making certain that the opening is cut to fit the sleeve and louvered grille correctly. Also make sure that the grille fits flush with the outside finished surface of your home. The grilles and trim may then be painted a color to match the finish of your home, making them hardly noticeable. Low shrubs planted in front of the grilles will make them even less conspicuous, but you must provide enough clearance for intake and exhaust air.

INSTALLATION INSTRUCTIONS

Each unit consists of several basic components, the number of which depends on the manufacturer. In most cases these components are all that is necessary for complete installation with the exception of some hand tools and a few miscellaneous items that can usually be purchased locally.

Before starting the installation, review the wall thicknesses and make certain that your measurements have been accurate, as they will dictate the method of installing the base unit. Where the walls are thicker than 9¼ in., or where built-in radiator enclosures are desired or already provided, usable floor space can be saved and the appearance of the installation enhanced if the base cabinet is partially recessed in the wall. However, if your walls are less than 9¼ in. thick, the rear extension may project beyond the outside face of the wall, as shown in Fig. 2-15. Obviously, this is undesirable, and the unit should be made flush with the outside wall by using a filler piece as illustrated in Fig. 2-15. If it is necessary to use a filler strip, add studs between the cabinet and the wall so that the cabinet can be fastened securely.

Start your wall openings following the rough-in dimensions packed with your unit. If you are cutting through wooden studs of an existing building,

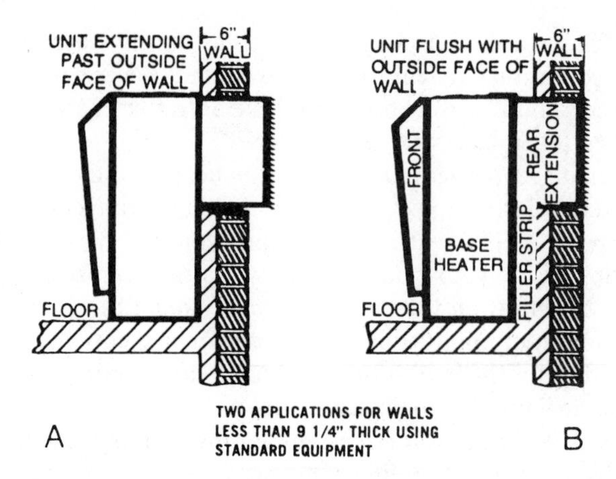

Figure 2–15. (a) On thin walls, the rear extension may project beyond the outside face of the wall, giving an undesirable appearance; (b) the projection is solved by using a filler piece to make the unit flush with the outside wall. (*Courtesy of Craftsman Book Company*)

frame the opening before installing the sleeve through the wall. Most units are designed to fit exactly in a given number of block or brick courses, so if you are installing in a masonry wall, all you will probably need to do is remove the required number of blocks or bricks and insert the sleeve. Once the unit is in place, the equipment itself will provide all the support necessary, so don't worry about angle-iron lintels.

Dimensions for locating the wall opening for typical through-wall heating and cooling units are shown in Fig. 2–16. Figure 2–17 gives useful suggestions for cutting the wall opening and sealing it once the sleeve is in place. The opening should have a neat finish, be weatherproofed, and the bottom of the opening should be pitched toward the outside approximately ¼ in. per foot for drainage. At this time you should make preparations for the electrical wiring and if you are using hot-water heat, for water pipes.

Before installing the base section and rear extension, apply nonhardening caulking compound around the inside of the wall opening as described in Fig. 2–17. Then apply caulking compound around the outside of the base rear extension, to about 1 in. from the rear edge of the cabinet body. Caulk the top and sides only at this time, unless the bottom edge of the masonry is rough. If this is the case, the bottom edge may be caulked also, but be sure to leave two or three 1-in. gaps (weep holes) in the caulking for moisture drainage.

Next, place the base heater rear extension in the wall opening and slide it to the desired location. Shim the bottom of the cabinet, if necessary, to make the rear extension approximately 1.4 in. low (below horizontal) at the end nearest the outside face of the wall. This is important, as it prevents condensed moisture from running toward the inside of your house.

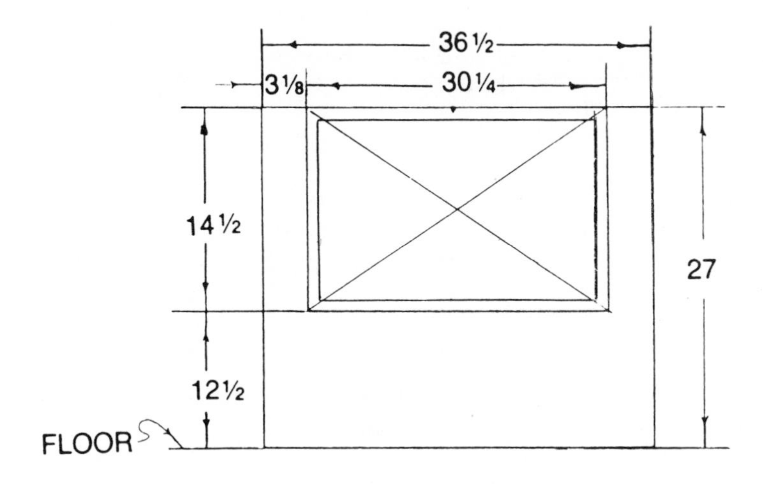

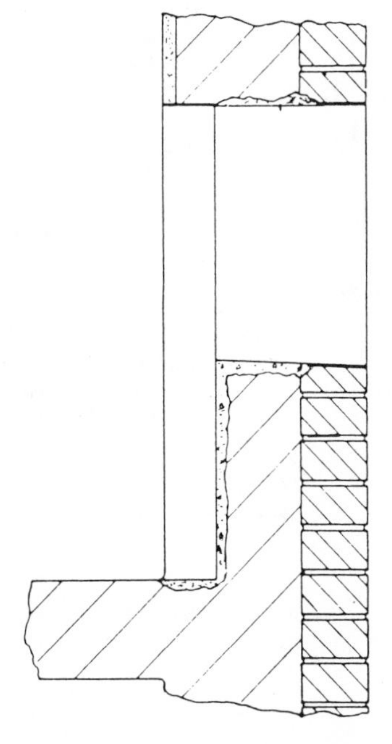

Figure 2–16. Wall section and dimensions for installing a recessed-cabinet unit. (*Courtesy of Craftsman Book Company*)

Secure the base unit to the floor and wall using suitable fasteners and check the caulking around the entire rear extension at the outside face of the wall. If the bottom face of the wall has been caulked, check the weep holes to make certain that they have not been clogged by the caulking.

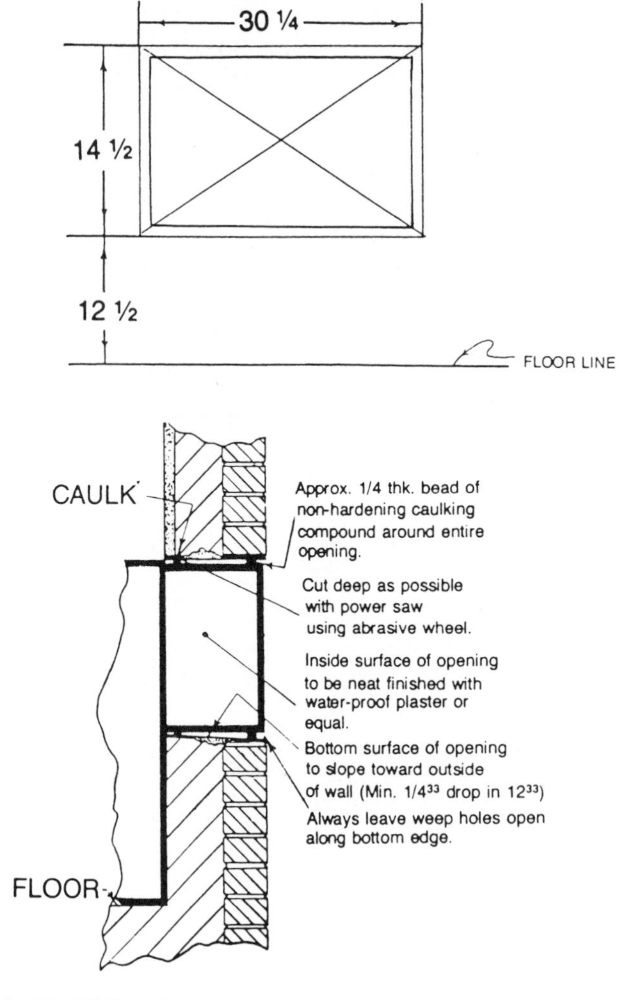

Figure 2–17. Wall sections and dimensions for installing a typical nonrecessed unit. Finishing and cutting instructions also apply to recessed units. (*Courtesy of Craftsman Book Company*)

PIPING FOR STEAM OR HOT WATER

The steam or water-heating coil in most through-wall units may be removed in order to make connections by removing two bolts at each end of the coil. However, the coil should be securely bolted in place and properly pitched before the final pipe connections are made. Either end of the coil may be used as the inlet or the return. The use of ½-in.-I.D. (⅝-in.-O.D.) copper tubing is recommended for connecting the heating coil to the pipe.

Once the heating coil has been pitched and bolted securely in place, the pipe connections may be completed as shown in Fig. 2–18. Note that holes are provided in both the back and bottom of the cabinet for heat piping. If the existing piping cannot be arranged to utilize these holes, others may be drilled in the cabinet to suit your particular needs.

For the type of unit shown in Fig. 2–18, you will need two ¾-in. × 2½-in. black iron nipples (item 3), one ¾-in. I.P.S. × 5.8-in. O.D. female copper adapter, and two lengths of ⅝-in. soft copper tubing long enough to reach from the heating coil to the existing piping. A shutoff valve is recommended on the inlet pipe to the coil, mounted as shown in Fig. 2–18 (item 8). In the case of hot-water piping, be sure to include an air vent (item 26). Other items will include a ½-in. balancing fitting (item 9) and one ½-in. × ½-in. × ¾-in. I.P.S. iron reducing tee. These items are not normally furnished with the units and will have to be purchased at your local hardware store or plumbing supply house.

With the piping out of the way, slide the blower assembly back into the cabinet extension (if it had to be removed) and reconnect the wires to the thermostat and junction box. However, do not install the mounting plate at this time.

The cooling chassis is now ready to be installed, but it is going to require

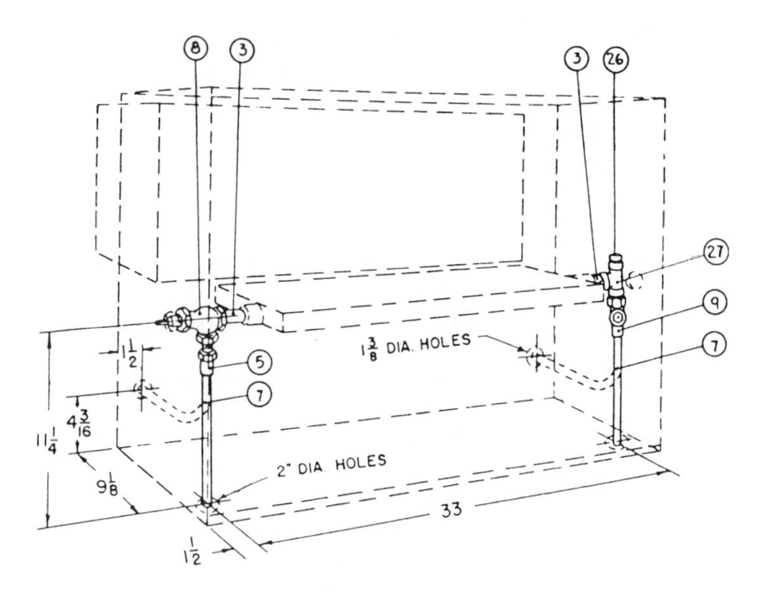

Figure 2–18. Pipe connections for a typical through-wall heating and cooling unit. (*Courtesy of Craftsman Book Company*)

some help, as it weighs too much for one person to handle. Uncrate the cooling chassis as close as possible to the base unit in which it is to be installed. With your hand, spin the condenser blower wheel to be sure that it has not become loose due to rough handling in shipment. If it has been loosened or the wheels rub against the blower housing, center the setscrew over the flat part of the motor shaft and retighten it.

Slide the cooling chassis all the way into the base section, but be extremely careful not to lift or pull it by any of the copper tubing forming the refrigeration circuit. When properly in place, reinstall the mounting plate.

Before proceeding any further, inspect the base convector rear extension to make sure that it is sealed and insulated properly. If any light shows through, or if there is any way that air might leak between the base section and the wall, caulk and insulate as necessary. This is very important since gaps will allow unconditioned outside air to infiltrate. In addition, in extremely cold weather gaps could cause steam traps or hot-water lines to freeze.

ELECTRICAL WIRING

A separate electrical circuit should be used for each conditioner; each should be provided with a disconnect device at the unit location. The feeder wire and overcurrent protection should be sized in accordance with the *National Electrical Code®*.

Always check the nameplate ratings on the cooling chassis (and on the heating coil if you are using an electric coil). In nearly all cases, the supply voltage will be 240 volts (V), but the amperage rating will vary with the size of the unit. If electric heat is used, the heating element load will be larger than the cooling load; since both operations will never run simultaneously, the wire size and overcurrent protection may be sized for the larger of the two, not for their sum.

For example, if the nameplate rating on your unit is 7.5 amperes (A) at 230 V for the cooling chassis and you are using hot water for heat, 7.5 A is the total load. Therefore, the wire would be sized as follows:

$$7.5 \text{ (amperes)} \times 1.25 \text{ (safety factor)} = 9.37 \text{ A}$$

Referring to the wire table (Table 2–1), No. 14 AWG is the smallest wire listed and is rated to carry 15 A. This is the size of wire to use. Conventional circuit breakers are not rated smaller than 15 A, so a 15-A breaker will be used.

But suppose that the nameplate stated an amperage of 14.9 A. Again, applying the safety factor, we have 14.9 × 1.25 = 18.6 A. Checking in the table, it is found that No. 12 AWG wire will be used and, in turn, provide an overcurrent device rated at 20 A.

TABLE 2–1. CURRENT-CARRYING CAPACITY OF COPPER WIRE

Wire Size	Ampere Rating
14	15
12	20
10	30
8	40
6	55

If an all-electric through-wall unit is used, a typical nameplate rating could state:

Cooling amperes: 7.5

Heating amperes: 12.5

In this case, the wire and overcurrent protection would be sized for the larger of the two loads, the heating load.

Besides the overcurrent protection at the electric panel, a separate disconnect device at the unit itself will be needed. This can be accomplished in several ways:

1. Have a remote switch in the power line to the unit.
2. Install a double-pole, single-throw (ON-OFF) switch of the proper rating in the cover of the junction box provided.
3. Install a 240-V receptacle in the junction box and attach a plug to the unit's power cord.

In all cases, be absolutely certain to ground the base cabinet to guard against harmful electric shock should a fault occur in the wiring. If armored cable is used for the circuit, the connection to the chassis will also provide a ground, provided that the cable is properly grounded at the breaker panel. If nonmetallic cable with ground wire is used, the bare ground wire should be securely fastened under some screw in the chassis or base section. Most units will have a special grounding screw (painted green) for this purpose.

With the wiring completed, install the filter and the cabinet front. Press the control button marked OFF, and turn on the electric power supply. If using a steam or hot-water heating coil, make sure that the inlet valve is turned off. Then press the button marked COOL, and turn the thermostat knob clockwise to the extreme COOLER position. The compressor will start immediately if the heating coil is cool, and in a few minutes the unit should be discharging cooled air.

The compressor may not start if the room is very cold; it may be necessary to submerge the thermostat bulb in warm water to get the compressor to start. Now turn the thermostat knob to the extreme WARMER position; the compressor should stop unless the room is very warm.

Now test the heating cycle. Press the button marked OFF and wait a few minutes before pressing the one marked HEAT, leaving the thermostat at the extreme WARMER position. The circulating-air fans will run at low speed—for normal heating—if the room is not too hot.

If you install an all-electric model, the heating element will also be energized and warm air will be felt in a few seconds. If the unit uses steam or hot water, you won't feel any significant change in temperature of the air unless the furnace and circulating pumps are operating. But if the blower fan operates, the unit itself is working properly. Press the button marked OFF and all motors should stop. The first unit has now been completed.

Continue this procedure for the remaining units until all are installed. Then paint the outside louvers to match the finish of the house. The inside unit can also be painted to match any room decor, including wood-grain finishes, or you might prefer a wooden enclosure, but be sure to leave an opening for the air discharge and return.

3

OIL-FIRED FURNACES

The oil-fired furnace is probably the most popular type of central heating unit in use in the United States at this time. Traditionally, two types have been used: the gravity warm-air system and the forced warm-air system. However, the former (gravity-system) is all but obsolete and is seldom found any more except in older homes.

Oil-fired, forced warm-air systems are by far the most popular type of central heating unit used in residential applications. Although more expensive than gravity systems, their advantages far outweigh the cost. In this type of heating system, circulation of the heated air is controlled by a fan or blower, and therefore does not necessarily have to be installed in the basement. A forced-air system can be installed in the garage, closet, utility room, in the attic, or practically any other location within a reasonable distance from the living area of the home. Other advantages include: the air can be filtered, ducts can be smaller, resisters can be placed in the walls or ceiling, ducts and pipes may be longer, and the entire system can be operated completely automatically. Another great advantage is that the system can easily be converted or adapted to a central air-conditioning system.

In an oil-fired, forced-air heating system, the oil flame heats up the air inside the burner compartment of the furnace. This warmed air is then pushed along by a blower fan, through sheet metal ductwork which opens to the various areas of the home. Return air ducts then carry this air back to the furnace to be reheated, usually after being mixed with some fresh air to keep it from becoming stale.

The mechanism for turning an oil-fired furnace on is relatively simple, usually being a conventional toggle switch with a red plate located at the top of the basement stairs in most cases. However, the complete control system is a bit more complicated with its various interlocking devices, which are designed to prevent accidental fire or explosion and many other safety features.

The high-limit control, for example, is a furnace thermostat that prevents overheating. In most cases this control is set just high enough to ensure proper temperatures without allowing the heater to overheat. With a forced-air system, the high-limit controls are usually set to start the fan when the furnace warms up, and to stop it when the heating unit cools down or when the area has reached the desired temperature. All must be adjusted to suit local weather conditions and the living characteristics of the home occupants.

In general, the thermostat in the room or area notifies the furnace when heat is needed. Some are equipped with timing devices to change the demanded temperature automatically, day or night. All settings are usually adjustable for prevailing conditions.

Other controls in the complete system ensure that the various operations, that is, when the fan should come on, and so on will take place in the proper order. Since most heating controls operate on low voltage, the control to the larger electrical items must be run through a relay. The relay is a magnetic switch that operates on a small current to handle the current for the oil burner motor and the ignition transformer. In other words, the contactors in the magnetic relay are controlled by low-voltage circuits, but when the contacts "snap in," it allows full household voltage (either 120 or 240 V) to pass on to the various electrical items used in the heating system.

Oil burner controls allow electricity to pass through the motor and ignition transformer and shut them off in the right order. They will also stop the motor if the oil does not ignite or if the flame goes out. This is done by a stack thermostat built into the relay. This whole assembly can be mounted inside the smoke pipe near the unit or on the wall. Without the protection of a stack thermostat, a gun-type or rotary-type oil burner could possibly flood the basement with fuel oil should it fail to ignite. Using such a control, however, allows the motor to run only a short time if the burner fails to ignite; then it opens the motor circuit and keeps it open until reset by hand.

In its very basic form, a schematic diagram of an oil furnace control system is shown in Fig. 3–1. Each component in this system is designed to either start, stop, regulate, or protect the heating system and its related components. This drawing shows the basic parts, consisting of the room thermostat (T); the transformer (C), which steps down the 120-V line voltage to about 24 V; the relay (R), which is an electrically controlled switch; the motor (M), which operates the pump that brings oil from the storage tank and forces it through an atomizer nozzle in the furnace; the ignition transformer (I), which steps up the 120-V ac line to about 10,000 V; and the spark gap (S), across which the high voltage jumps to form an intense electrical flame.

The thermostat (T) is a temperature-sensitive switch which is used to

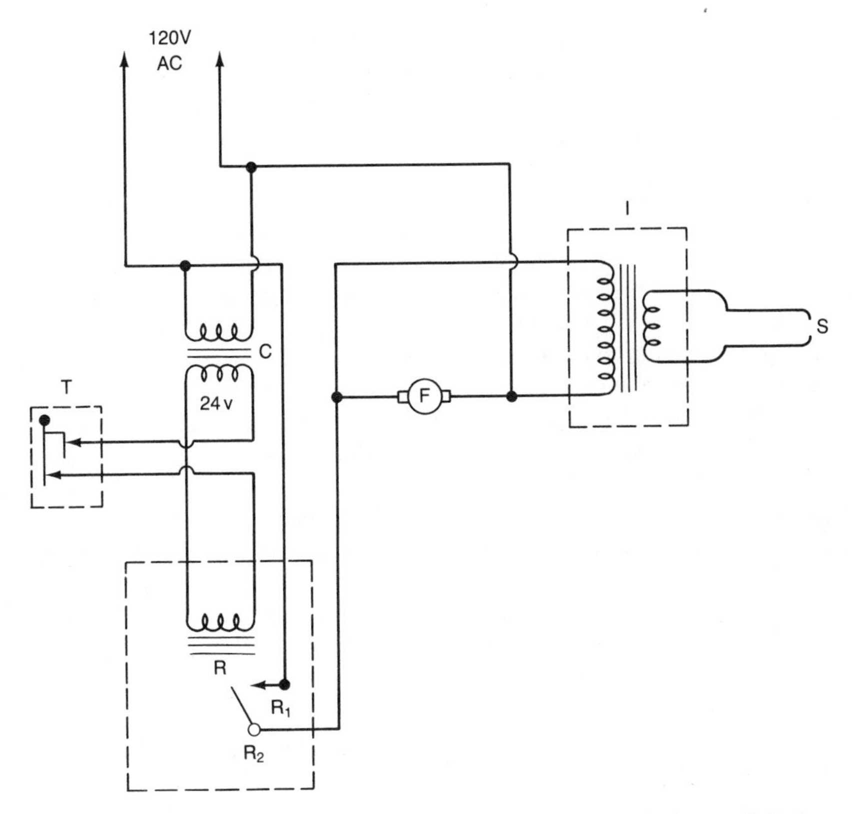

Figure 3–1. Schematic diagram of a simplified control circuit for an oil-fired heating system.

sense the temperature of the area in which it is placed. When properly set, the heating system becomes activated when the room temperature becomes too low, since the thermostat is designed to sense the room temperature and then to signal the heating unit for the proper response. When the former condition exists, the two contacts of the thermostat close, which signals the relay (R) that the area is "calling for heat." This 24-V current going through the coils of the relay energizes the relay, pulling contacts R1 and R2 closed, which in turn allows the 120-V current to flow to the motor terminals, which start the flow of oil. At this same time, the ignition transformer (I) is energized. The pumped oil is vaporized into the furnace and this vapor is ignited by the sparking across the gap of the spark gap (S). Once the oil starts to burn, the flame will maintain itself without needing the spark of the spark gap.

The flame heats the air, and when of sufficient temperature, the furnace blower comes on to circulate the air throughout the house. When the air in the vicinity of the thermostat reaches the desired temperature, the contacts in the thermostat open, breaking the relay circuit and causing the relay contacts to open; this in turn cuts off the motor and the oil supply, and the flame dies out. In some cases, the blower fan is designed with a delayed switch that allows the fan to continue to run for a while once the burner flame goes out.

A more detailed drawing of a heating control circuit might include an overcurrent protection device to protect the entire circuit. Furthermore, it will probably have an emergency cutoff switch, located in a convenient place so that the entire system may be shut down for various reasons. Then, other protective devices, such as the stack control mentioned previously, will be included. In a normal condition, oil flames and hot exhaust gases start up the stack to the chimney. During this movement, heat-sensitive arms in the stack, which are connected to electrical contacts, begin to twist. If sufficient heat is present, the arms will twist to close the contacts and keep the heating system in operation. However, if after several seconds, the twisting is not sufficient to close the contacts, the circuit opens and shuts down the heating system. So this excellent protective device guards against any number of faults that can occur in the system: open or shorted circuit, fouled spark gap, and the like.

In some oil-fired heating systems, the furnace is also used to supply domestic hot water. A coil of heavy copper pipe carrying cold water is immersed in the water jacket of the boiler. The cold water is heated by contact with the boiler water and passes on to a storage tank. In a "tankless" system, the copper pipe is big enough to act as its own reservoir. In winter, when the furnace operates frequently, hot water is plentiful. In summer, an independent thermostat in the water jacket turns the burner on for short periods to heat the desired amount of water, but not high enough to bring on the space heating cycle.

FURNACE CONTROLS

A fully automatic oil-fired furnace control system performs the following basic functions:

1. Sensing element measures variations in humidity, pressure, and temperature.
2. The control mechanism converts these variations into energy that can be used by fan motors, motorized dampers, valves, and so on.
3. In the case of an electric or electronic control system, the connecting electric wiring transmits the energy to the various devices to be controlled.
4. The devices then use this energy to initiate some corrective action to satisfy the variation in temperature, pressure, and/or humidity.
5. When the variations are brought under control by the sensing elements, the sensing element signals the control mechanism or connecting means

to stop the corrective action, since the desired conditions have been satisfied.

Figure 3–2 shows a sketch of a wiring diagram of a blower motor in a fan-coil unit. This is typical of those that accompany the heating units, and

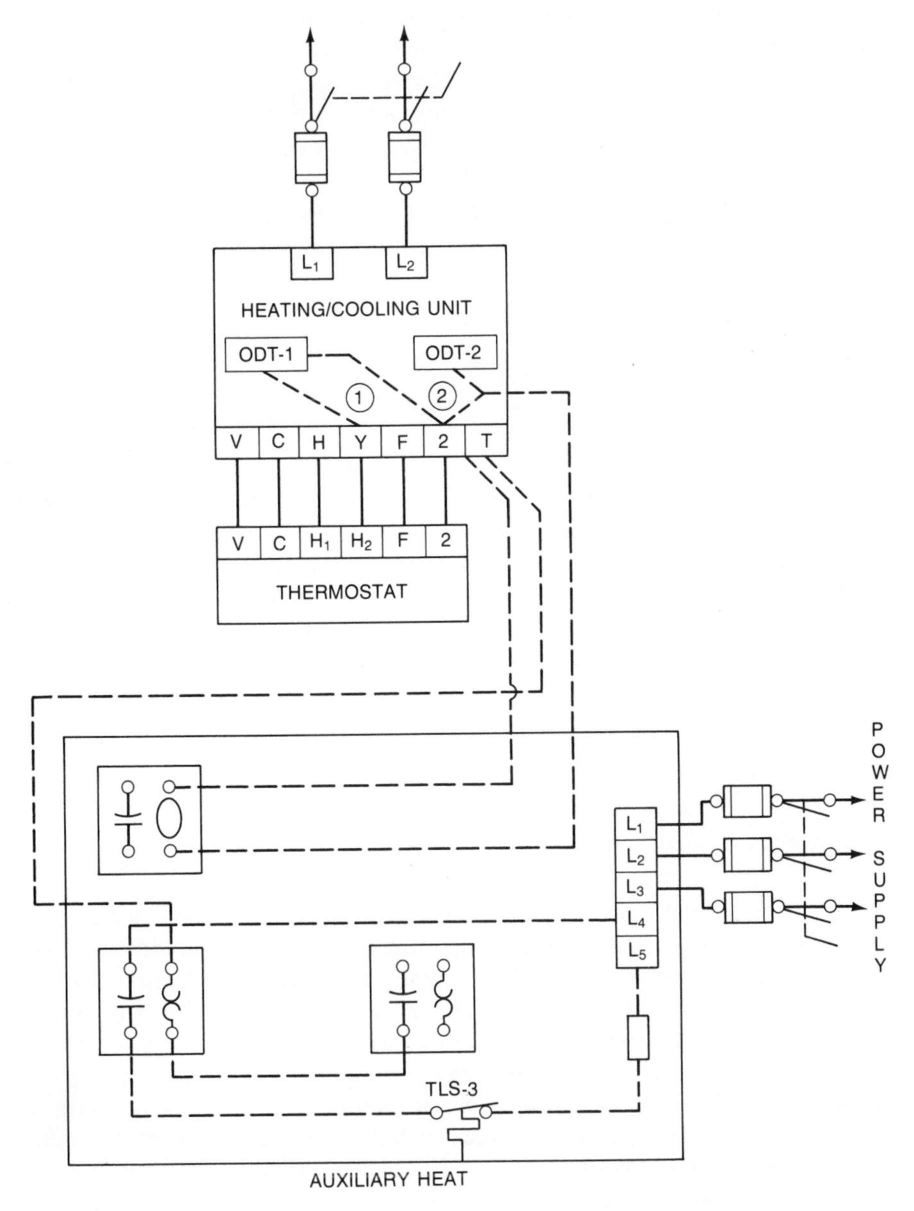

Figure 3–2. Field wiring diagram for a heating/cooling system.

are used mainly for connection purposes. However, such drawings are also excellent for troubleshooting purposes; that is, they may be used to trace each conductor in the control system to check for loose connections and the like.

A humidistat, when used in conjunction with a heating system, works similarly to a thermostat. A humidistat is designed to maintain a certain level of humidity in the circulating air. Most operate on the change in tension of hair springs, caused by variation in humidity. When the humidity is below the desired percentage, the humidistat activates a humidifier (usually installed within the ductwork), which ejects water vapor into the conditioned air. One type of humidistat operates in the following manner.

Fill Cycle: A solid-state filling cycle is independent of the humidifying operation and consists of the following components:

1. Solid-state circuit board
2. Water-level probes
3. Fill valve
4. Current transformer

During the fill cycle, the control transformer converts the line-voltage power supply to 24 V of control power. The fill-valve operation is controlled by a flush cycle, the stainless-steel water-level probes, and the solid-state circuit board. A constant water level is maintained except during the flush cycle.

Humidification Cycle: Almost instantaneous water evaporation is accomplished by the following components:

1. Heating element
2. Control relay
3. Control transformer

Whenever the humidistat senses a need for more humidity, it energizes the control relay and starts the heating element and the system blower (if not already in operation). When the humidistat is satisfied, it deenergizes the control relay, shutting off the heating element and the system blower (provided that the blower operation is not required by other controls). The power and water supply to the humidifier must be left on at all times. During summer or other periods where humidification is not required, the humidistat control is turned to OFF and the heating element will not operate. However, the solid-state fill controls will continue to keep the reservoir filled, ensuring a clean start for the next heating season.

Flush Cycle: The timed flush cycle removes residuals from the reservoir and consists of the following components:

1. Flush valve
2. Fill valve
3. Control transformer
4. Clock timer
5. Drain

During the flush cycle, the clock timer opens the flush valve; the solid-state fill controls open the fill valve; and six precisely located water-inlet jets swirl water into the reservoir, flushing residuals down the drain. The flushing continues for approximately 3 minutes; then the flush valve closes and the fill-cycle controls shut off the fill valve when the reservoir is refilled. At the factory the humidifier is adjusted to give one flush cycle for every 2 hours of heating-elements operation. However, field adjustment of the timer can give a choice of flush-cycle frequency from 2 to 6 hours. A flush cycle every 2 hours of ON time is recommended for most jobs. Where soft water is prevalent or where a water softener is included in the plumbing system, the adjustable timed flush cycles can be decreased to 6 hours of ON time.

The timer has a manual override control knob which when turned to the proper position will give a flush whenever desired. Turning this knob also gives a positive check of all humidifier operation functions.

Overflow Protection: The transparent reservoir has a separate overflow drain compartment. In the event of flush or fill-valve failure, water will run out this independent overflow drain compartment. The overflow safety probe is in the solid-state circuit and will shut off the fill valve in the event the drain line becomes plugged. The manually reset capillary-actuated limit control senses the heating-element temperature and shuts off the element when a temperature of 260° F is reached.

4

GAS-FIRED FURNACES

Gas heating devices can be divided into groups or classes, depending on the method used for supplying and controlling the heat generated by the gas burner. Manually controlled room heaters are used commonly in some installations, while completely automatic furnaces are used in other residential installations.

Manual room heaters are used because of comparatively low initial installation cost and economy in operation, since each room may be individually controlled. They are, as the name implies, nonautomatic and can supply heat to an area quickly by igniting the gas when needed. If only one or two rooms require heat during the day, the other heaters may be shut off until they are needed. However, when they are used continuously at a rather low temperature setting, much of their economy is lost.

Automatic gas room heaters and furnaces are the type that will more frequently be found to heat residential dwellings. Two types are in common use: the forced hot air type and the gas boiler for hot-water heating. Both are relatively efficient and have the capability of furnishing clean heat for practically any type of installation, both new and existing.

One of the most frequently encountered types of gas furnaces is the floor-insert type. Because of the nature of these units, servicing should be handled with care, observing all safety rules. To start, if the gas pilot has been on but not lighted, make certain that all lines and gas burning equipment has been purged by shutting off the main gas valve and waiting at least 30 minutes. The reason for this is that liquid propane gases are heavier than air and will not rise out of the exhaust pipe or chimney. Most automatic furnaces have an automatic shutoff should the pilot go out. However, it takes several seconds—if not minutes—for this safety feature to activate fully, and it is possible to get quite a bit of built-up gas in the base of the heater before the gas supply is completely shut off. Therefore, before any repairs are made, make certain that this excess gas is allowed to escape into the surrounding atmosphere to dilute it to a safe mixture.

The first item to check is the air filter. This should be replaced or at least thoroughly cleaned at least once each year, more frequently if necessary. Better heating and ventilation can be had if they are cleaned once every month.

Next turn to the burner, removing any accumulated carbon through the removable door at the front or side of the burner. Also remove all dirt and dust from inside the casing through the cleanout doors.

One item that often gets overlooked on gas furnaces is the smoke pipe. Since gas is so clean in comparison to coal and oil, few people ever think of inspecting the vent or smoke pipes. The best time to renew a vent or smoke pipe is during the summer when the furnace is not in use. However, before actually removing the pipe, remove any controls or check drafts that are attached to the pipe. Perhaps these will require special fittings to fit into a new pipe if one is required. If they can be replaced, try to remove the pipe and controls as a unit, laying them aside until a new pipe is installed. This gives a guide to installing the controls in the new pipe, as well as taking measures if any cutting must be done.

Smoke pipe is available at any heating or hardware store, and at least 24-gauge metal is recommended. Elbows and angles should be of the adjustable type and of the same gauge metal as the pipe itself.

The tools required to replace a smoke pipe are simple and should include an electric drill, sheet metal screws, screwdriver, circle-cutting tin snips, a trowel, and some freshly mixed mortar or furnace cement for sealing around the pipe in the thimble.

The existing pipe may be used as a guide, provided of course that it gave satisfactory service. Exact measurements may be taken from the existing pipe, transferred to the new pipe, and then appropriate action taken, that is, cuts, holes for controls, and so on. Depending on the amount of room available, it may sometimes be necessary to install the pipe in short sections and then add the draft control and the like afterward; or where a lot of room is available, some prefer to install the entire pipe in one section with all controls intact.

When removing any existing controls from the old pipe, carefully observe (and make notes) exactly how they are installed. Most are screwed in place with sheet-metal screws and brackets. Measurements are taken and then holes are drilled in the new pipe with an electric drill, although many mechanics make this with a scratch awl and hammer.

If the furnace is equipped with an electric damper control, the chains or other items used to regulate the damper must be installed exactly as they were previously. As a check, the thermostat setting may be moved, which should cause the draft door at the bottom of the furnace to close when the check draft door is open, and vice versa. However, this type of control is becoming obsolete and is mentioned here only in case you encountered one that is still in use. Most modern heating and cooling controls are of the electronic type, but will still activate small electric motors to control the draft, zones, and so on. These will also have to be checked.

Once the new pipe has been installed, it is now ready to be sealed with mortar, using about half sand and half portland cement or furnace cement. The seal should be airtight to have maximum efficiency. Before making the seal, however, make absolutely sure that the pipe extends into the chimney only as far as the inner edge of the flue tile or chimney opening.

The remaining items on a gas-fired furnace checklist include the following:

1. Thoroughly clean the entire furnace.
2. Oil any motors.
3. Check fan and other belts. Replace worn-out belts.
4. Check and replace any firebricks in furnace lining.
5. Blank off an unused heat pipe or leader.
6. Renew air filters.
7. Replace fuse plug.
8. Eliminate minor rattles in the furnace or ductwork.

Gas-fired heating systems are very similar to the oil-fired systems discussed in Chapter 3, except that gas-fired units are usually simpler because they do not require a pump motor or an ignition system. Instead of the pump motor, there is a magnetically operated gas valve, and for ignition there is a small, permanently lighted pilot flame in the flame box. However, safety features are also included in the controls of a gas-fired system to protect the system, components, and life and property.

Any failure of the gas supply causes a thermal safety valve to cool down and lock the main gas line shut. It cannot come on again until the pilot is relighted and allowed to reheat the safety valve. This protects the occupants of a home from becoming killed or injured due to a malfunction in the system. It is easy to imagine what would happen if the flame and pilot went out and the gas continued to flow in the area for any length of time. Any spark or lighted match could cause a terrible explosion.

Forced-air gas-fired furnaces are extremely popular in areas where natural gas is available because they are clean to operate, easy to maintain, and quick heating, in addition to being about the most economical to operate next to coal. Like other central units, gas-fired furnaces lend themselves to use in combination with a cooling system for summer comfort.

A typical schematic wiring diagram for a gas-fired furnace control circuit is shown in Fig. 4–1. In general, the overcurrent protection (CB) consists of a circuit breaker in the main load center. This protects the main 120-V line supplying power to the system for operation. Note that a main shutdown switch is connected in series with the line; this acts as an emergency cutoff means for repairs, fires, and the like. In other words, either of these two devices (the circuit breaker or the shutdown switch) will cause all contacts to open and deenergize the entire circuit. The 120-V lines connect to a transformer (C) which steps down the 120-V current to about 24 volts; the primary side of this transformer is 120 V and the secondary side is 24 V.

From the secondary side of the step-down transformer, the circuit runs in series to a room thermostat (T), switches 2 and 3, and to an electromagnetic valve (V). If this circuit is functioning properly, when the room calls for heat, the contacts of the thermostat close. However, before it can actuate the valve (V), which regulates the gas to the burners, switches 2 and 3 must be closed. The first (S2) is controlled by a pushrod which is part of a thermostatic element exposed to the pilot light in the gas chamber. If the pilot is functioning properly, the switch will remain closed so that current can flow through the switch. However, should the pilot fail at any time, the element cools off and quickly locks the switch in its open or OFF position, and current cannot flow in the circuit. If the latter condition exists, regardless of the thermostat signal, the system will not operate. Once the pilot is relighted, the element will again heat up and S2 will close to the ON position.

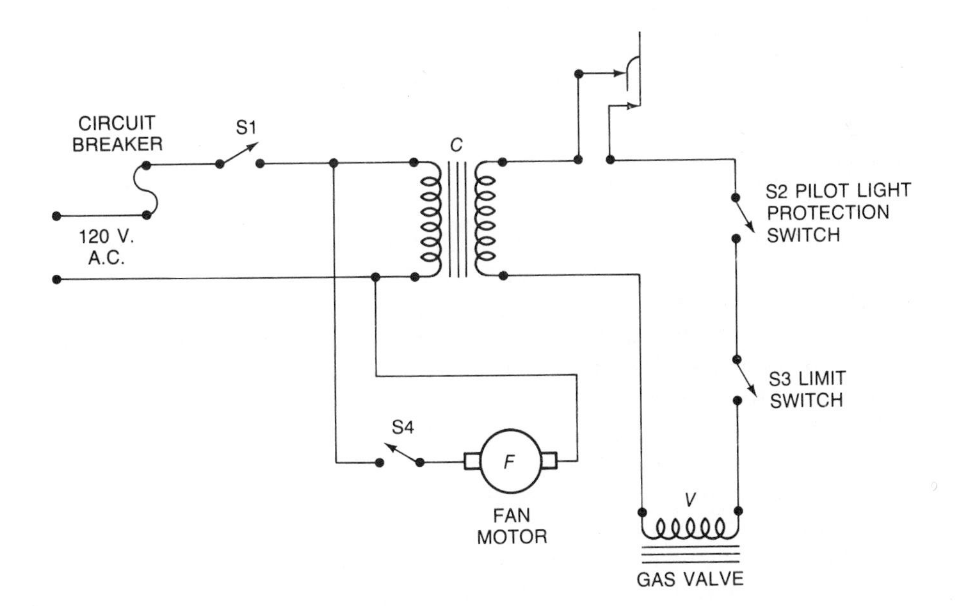

Figure 4–1. Schematic diagram of a typical gas-fired, forced-air furnace.

The other safety switch, S3, acts as a "limit" switch in that it normally acts in combination with a switch for the furnace blower. In the majority of cases, both of these switches (S2 and S4) will have their contacts closed, so that if the room thermostat calls for heat, current will flow through the circuit to operate the gas valve (V). As the gas is emitted through its port, the pilot light will ignite it and keep it burning. Since it takes a while for the air to be heated, it would not be desirable for the furnace blower fan to come on immediately, as it would send cold air throughout the home, making the area highly uncomfortable until the furnace heated the air to a suitable temperature. Instead, a thermostatic element in S3, working in conjunction with S4 (interlocked), keeps S4 open for about 5 minutes, while the air in the furnace becomes heated. After this period, S4 closes, energizes the fan, which in turn pushes warm air through the ductwork to be distributed throughout the house. After the desired temperature in the rooms have been reached, the thermostat contacts open, which eliminates the flow of electricity in the control circuit, that is, all but the fan blower. In most systems, the fan circuit remains on for a few minutes after the furnace quits while it empties the remaining warm air in the system. Then this switch (S4) opens and the fan stops. The entire system will then be shut down until the room thermostat once again calls for heat; then the cycle will begin again.

The limit switch (S3) further prevents the system from becoming overheated. Most limit switches are calibrated to some predetermined temperature. Should the room thermostat be set too high, or another malfunction in the system causes the heated air to become too hot for comfort or safety, the limit switch will automatically open, deenergizing the control circuit, which in turn will shut the main gas valve. However, since S4 is thermostatically controlled, the fan will continue to run momentarily until the system is cleared of hot air.

TROUBLESHOOTING

Most problems with gas-fired furnaces will develop in the control system, and most problems can therefore be determined with a few hand tools and an accurate volt-ohmmeter. Just be extremely careful when working with "live" electrical wires of any voltage; most systems will be located in the basement of the home, where a damp concrete floor does not add any safety features to working around electricity. If possible, always work from a wooden stepladder or at least use a dry wooden board to stand on during the testing procedure. Also, avoid touching any water pipes during the testing, and by all means, never touch any open wires or contacts.

When trouble develops with a gas-fired furnace, look for a schematic diagram, which is usually attached to the furnace itself, perhaps on the back of an access door. Familiarize yourself with the electrical circuitry of the system before attempting any troubleshooting procedures, especially with the color coding of the wires or conductors.

Unless experience dictates otherwise, the most logical place to begin is at the source of the electrical supply, the main distribution panel, where overcurrent protection is provided. Using the test probes of the voltmeter, place one lead on the neutral wire of the circuit and the other on the load side of the circuit breaker. If no reading is present, chances are that the circuit breaker has tripped, and this should be held in the OFF position for a second or two and then flipped back to the ON position. Take another reading. If the circuit shows to be ''hot'' but the furnace still does not function properly, continue the check. The reading should be somewhere between 110 and 123 V.

The next point is the emergency shutoff switch. With one of the probes on the tester making contact with the neutral or ground wire, touch the other probe to both sides of the switch contacts. Again, a reading of between 110 and 123 V should be present if the switch is closed (ON position). Continue checking the voltages at each connection, but remember, once you get to the secondary side of the step-down transformer, the voltage should be reduced to about 24 V instead of the normal 120 V.

The schematic diagram that accompanies the unit should give the proper voltages at the various components. If a reading cannot be obtained as it should be, the first thing to look for is loose connections rather than for defective parts. Try tightening all screws and other connections; then look for broken wires. Finally, if all else fails, start looking for defective parts, but again, these can usually be found by using the volt-ohmmeter.

COMBUSTION AIR SUPPLY

Adequate facilities for providing air for combustion and ventilation must be provided for all gas furnaces. Where appliances are installed within a confined space and combustion air is taken from within the heated space, the air supply must be through two permanent openings of equal area, one located approximately 6 in. below the ceiling and one approximately 6 in. above the floor. The total free area of each opening should be equal to 1 in.^{2} per 1000 Btu/h to the total input rating of all appliances in the enclosure and in no case less than 100 in.^{2}. Where appliances are installed in a confined space within a building of unusually tight construction, air for combustion must be obtained from outdoors or from spaces or ducts freely communicating with the outdoors. Under these conditions, two openings of approximately equal area must be provided, each with a total free area of not less than 1 in.^{2} per 4000 Btu/h of total input rating of all appliances in the enclosure. If horizontal ducts are used, each opening must have a free area of not less than 1 in.^{2} per 2000 Btu/h of the total input rating of all appliances in the enclosure.

Where appliances are installed in an unconfined space in buildings of conventional frame, brick, or stone construction, infiltration normally is adequate to provide air for combustion, ventilation, and draft hood dilution. If the unconfined space is within a building of unusually tight construction, a supply of combustion, ventilation, and draft hood dilution air must be obtained from the outdoors. This must be through permanent openings having a total free area of not less than 1 $in.^2$ per 5000 Btu/h of the total appliances input rating.

MAIN BURNER PRIMARY AIR ADJUSTMENT

After a gas-fired furnace has been in operation for a short period of time each season, inspect the flames, which should have soft blue cones. If the flame does not appear to be correct, the burner should be adjusted by the following steps:

1. Close the air shutters until yellow tips appear on the flames.
2. Open the air shutter until the yellow tips just do disappear and the flames have soft blue cones.
3. Repeat this procedure on all burners.
4. Lock the air shutter by means of the lock screw.

BETTER GAS-FIRED FURNACE HEATING

1. Keep the air filters clean as this will enable the heating system to operate more efficiently and provide better heating, more economically.
2. Arrange furniture and drapes so that the supply air registers and the return air grilles are unobstructed.
3. Close doors and windows. This will reduce the heating load on the system and increase the efficiency of the furnace.
4. Avoid excessive use of kitchen exhaust fans.
5. Do not permit the heat generated by television, lamps, or radios to influence the thermostat operation.
6. It may be desirable to extinguish the pilot burner during the summer months, as some authorities have calculated a savings in excess of 3000 ft^3 of gas without pilot operation during this period. To do so, turn the main gas valve to the OFF position. This will prevent any further flow of gas to the pilot burner. Review the lighting instructions, however,

prior to establishing a pilot flame at the beginning of the next heating season.

7. Except with the possibility of the mounting platform, keep all combustible articles 3 ft from the furnace draft diverter and vent stack.
8. Replace all blower doors and compartment covers after servicing the furnace. Do not operate the unit without all panels and doors securely in place. If it is desired to operate the system with constant air circulation, it is advisable to seek advice from the manufacturer.

LUBRICATION

The blower motor bearings on most modern gas-fired furnaces are prelubricated by the motor manufacturer and may not require attention for an indefinite period of time. However, the following is recommended:

1. Motors having no oiling ports have been prelubricated and sealed. No further lubrication should be required, but in case of bearing problems, the blower and the motor end bells can be disassembled and the bearings relubricated by a qualified technician.
2. Motors having oiling ports require from 10 to 20 drops of electric motor oil or an SE grade of nondetergent SAE-10 or 20 motor oil to each bearing every 2 years for somewhat continuous duty, or at least every 5 years for light duty. Take care not to over-oil, because excessive lubrication can damage the motor.

Regardless of the motor type, all should be cleaned periodically to prevent the possibility of overheating due to an accumulation of dust and dirt on the windings or on the motor exterior.

TROUBLESHOOTING HINTS

Problem 1: *The heat is insufficient, but the burner and blower are operating.*

Remedy

1. Increase the temperature setting on the thermostat.
2. Check the return air filters and change, if necessary.
3. Recheck to assure that all supply registers and diffusers are open.
4. Check the closing of all doors and windows.
5. Check that the blower compartment doors are in place.
6. Call a servicing contractor.

Problem 2: *The pilot burner is on, but the main burner is off.*

Remedy

1. Raise the thermostat setting above room temperature.
2. Check the power supply fuses.
3. Refer to the lighting instructions.
4. Check for ON position of gas valve.
5. Call a servicing contractor.

Problem 3: *The main burner operates, but the blower ON-OFF control is fast (recycling).*

Remedy

1. Check the cleanliness of the air filters.
2. Call a servicing contractor.

5

ELECTRIC HEATING SYSTEMS

Less than 25 years ago, electric heating units were used only for supplemental heat in small, seldom-used areas of the home, such as a laundry room or workshop, or in vacation homes on chilly autumn nights. Today, however, electric heat is used extensively in both new and renovated homes.

In addition to the fact that electricity is the cleanest fuel available, electric heat is usually the least expensive to install and maintain. Individual room heaters are very inexpensive compared to furnaces and ductwork required in oil or gas forced-air systems, no chimney is required, no utility room is necessary since there is no furnace or boiler, and the installation time and labor are less. Combine all of these features and we have a heating system that ranks with the best.

There are several types of electric heating units available (Fig. 5–1) and a description of each will help you decide which will best suit your needs, and also help in the repair and maintenance of such units.

ELECTRIC BASEBOARD HEATERS

As the name implies, this type of heater is mounted on the floor along the baseboard, preferably on outside walls under windows for the most efficient operation. It is absolutely noiseless in operation and is the type most often used for heating residential occupancies. The ease in which each room or area may be controlled separately is another great advantage of this heater. Living

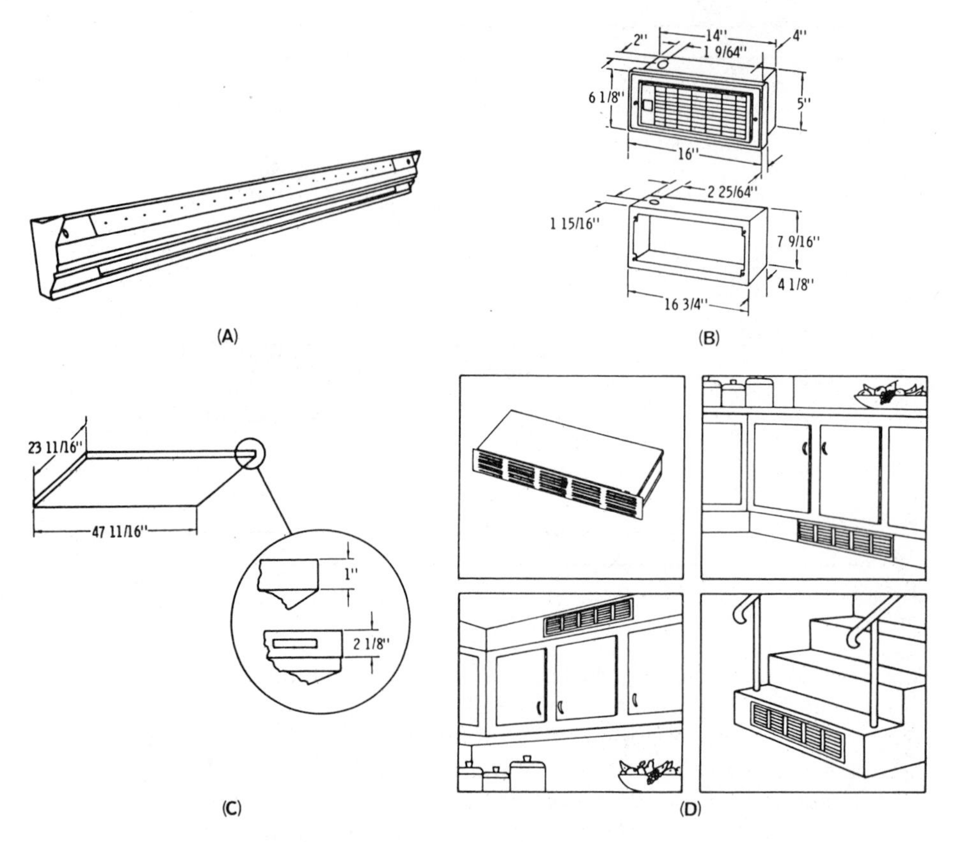

Figure 5-1. Several types of electric heating units: (a) electric baseboard, (b) fan-forced wall units, (c) radiant ceiling panel, and (d) four applications of kick-space heaters. *(Courtesy of the author)*

areas can be heated to, say, 70°F; bedrooms lowered to, say, 55°F, for sleeping comfort; and unused areas may be turned off completely.

Electric baseboard units may be mounted on practically any surface (wood, plaster, drywall, etc.), but if polystyrene foam insulation is used near the unit, a ¾-in. (minimum) ventilated spacer strip must be used between the heater and the wall. In such cases, the heater should also be elevated above the floor or rug to allow ventilation to flow from the floor upward over the total heater space.

One complaint received over the years about this type of heater has been wall discoloration directly above the heating units. When this problem occurred, the reason was almost always traced to one or more of the following:

1. High wattage per square foot of heating element
2. Heavy smoking by occupants
3. Poor housekeeping

RADIANT CEILING HEATERS

This type of heater is often used in bathrooms and similar areas so that the entire room does not have to be overheated to meet the need for extra warmth after a bath or shower. They are also used in larger areas, as a garage or basement, or for spot-warming a person standing at a workbench.

Most of these units are rated from 800 to 1500 watts (W) and normally operate on 120-V circuits. As with most electric heating units, they may be controlled by a remote thermostat, but since they are usually used for supplemental heat, a conventional wall switch is usually used. They are quickly and easily mounted on an outlet box in much the same way as conventional lighting fixtures. In fact, where very low wattage is used, ceiling heaters may often be installed by merely replacing the ceiling lighting fixture with a heater.

RADIANT HEATING PANELS

Radiant heating panels are commonly manufactured in 2-ft × 4-ft sizes and are rated at 500 W. They may be located on ceiling or walls to provide radiant heat which spreads evenly through the room. Each room may be controlled by its own thermostat. Since this type of heater may be mounted on the ceiling, their use allows complete freedom of room decor, furniture placement, and drapery arrangement. Most are finished in beige to blend in with nearly any room or furniture color.

Units mounted on the ceiling give the best results when located parallel to, and approximately 2 ft from, the outside wall. However, this type of unit may also be mounted on walls.

ELECTRIC INFRARED HEATERS

Rays from infrared heaters do not heat the air through which they travel. Rather, they heat only persons and certain objects which they strike. Therefore, infrared heaters are designed to deliver heat into controlled areas for the efficient warming of people and surfaces both indoors and outdoors (such as to heat persons on a patio on a chilly night, or around the perimeter of an outdoor swimming pool). This type of heater is excellent for heating a person standing at a workbench without heating the entire room; melting snow from steps or porches, sunlike heat over outdoor areas, and similar applications.

Some of the major advantages of infrared heat are:

1. No warm-up period is required. Heat is immediate.
2. Heat rays are confined to the desired areas.

3. They are easy to install, as no ducts, vents, and so on, are required.
4. The infrared quartz lamps provide some light in addition to heat.

When installing this type of heating unit, never mount the heater closer than 24 in. from vertical walls unless the specific heating unit is designed for closer installation. Read the manufacturer's directions carefully.

FORCED-AIR WALL HEATERS

This type of heater is designed to bring quick heat into an area where the sound of a quiet fan will not be disturbing. Some are very noisy. Most of these units are equipped with a built-in thermostat with a sensor mounted in the intake airstream. Some types are available for mounting on high walls or even ceilings, but the additional force required to move the air to a usable area produces even more noise.

FLOOR INSERT CONVECTION HEATERS

This type of electric heater requires no wall space, as they fit into the floor. They are best suited for placement beneath conventional or sliding glass doors to form an effective draft barrier. All are equipped with safety devices such as a thermal cutout to disconnect the heating element automatically in the event that normal operating temperatures are exceeded.

Floor insert convector heaters may be installed in both old and new homes by cutting through the floor, inserting the metal housing, and wiring according to the manufacturer's instructions. A heavy-gauge floor grille then fits over the entire unit.

ELECTRIC KICK-SPACE HEATERS

Modern kitchens contain so many appliances and so much cabinet space for the convenience of the housewife that there often is not room to install electric heaters except on the ceiling. Therefore, a kick-space heater was added to the lines of electric heating manufacturers to overcome this problem.

For the most comfort, kick-space heaters should not be installed in such a manner that warm air blows directly on occupants' feet. Ideally, the air discharge should be directed along the outside wall adjacent to normal working areas, not directly under the sink.

RADIANT HEATING CABLE

Radiant heating cable provides an enormous heating surface over the ceiling or concrete floor so that the system need not be raised to a high temperature. Rather, gentle warmth radiates downward (in the case of ceiling-mounted cable) or upward (in the case of floor-mounted cable), heating the entire room or area evenly.

There is virtually no maintenance with a radiant heating system, as there are no moving parts and the entire heating system is invisible—except for the thermostat.

COMBINATION HEATING AND COOLING UNITS

One way to have individual control of each room or area in your home—as far as heating and cooling is concerned—is to install through-wall heating and cooling units. Such a system gives the occupants complete control of their environment with a room-by-room choice of either heating or cooling at any time of year—at any temperature they desire. Operating costs are lower than many other systems due to the high efficiency of room-by-room control. Another advantage is that if a unit should fail, the defective chassis can be replaced immediately or taken to a shop for repair without shutting down the remaining units in the home.

When selecting any electric heating units, obtain plenty of literature from suppliers and manufacturers before settling on any one type. In most cases you are going to get what you pay for, but shop around at different suppliers before ordering the equipment.

Delivery of any of these units may take some time, so once the brand, size, and supplier have been selected, order in plenty of time before the unit is actually needed.

ELECTRIC FURNACES

Electric furnaces are becoming more popular, although somewhat surpassed by the all-electric heat pump. Most are very compact, versatile units designed for either wall, ceiling, or closet mounting. The vertical model can be flush-mounted in a wall or shelf mounted in a closet, while the horizontal design (Fig. 5-2) can be furred into a ceiling (flush or recessed).

Central heating systems of the electrically energized type distribute heat from a centrally located source by means of circulating air or water. Compact electric boilers can be mounted on the wall of a basement, utility room, or closet and will furnish hot water to convectors or to embedded pipes, with the

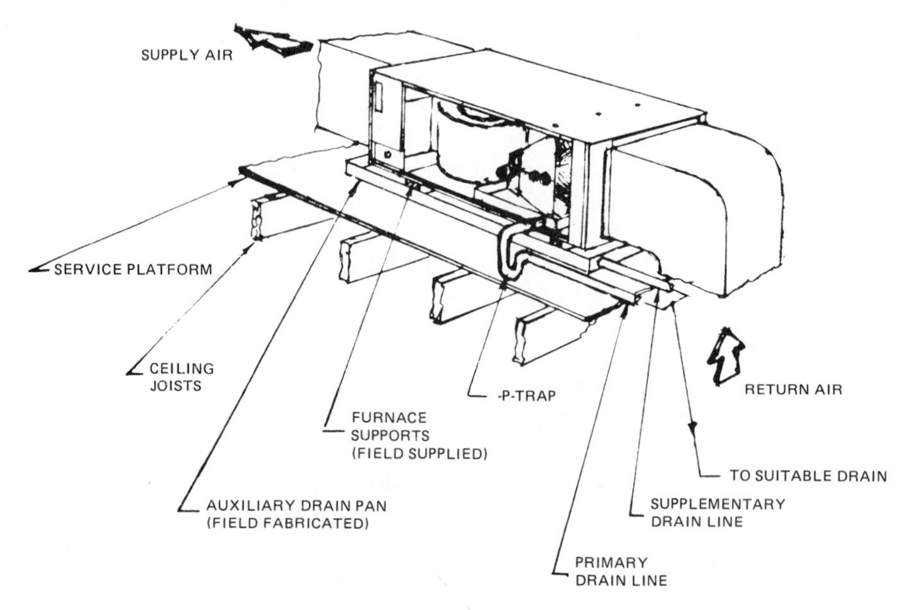

Figure 5–2. Horizontal application of an electric furnace. (*Courtesy of Square D Company*)

necessary controls and circuit protection. Immersion heaters may be stepped in one at a time to provide heat capacity to match heat loss.

The majority of electric furnaces are commonly available in sizes up to 24 kilowatts (kw) for residential use. The larger boilers with proper controls can take advantage of lower off-peak electricity rates, where they prevail, by heating water during off-peak periods, storing it in insulated tanks, and circulating it to convectors or radiators to provide heat as needed.

ELECTRIC HEATING CONTROLS

Electric heating units of all types may be effectively controlled by either line-voltage or low-voltage thermostats and related controls. In general, a line-voltage thermostat acts simply as an automatic switch to turn the heater ON and OFF to maintain temperature in the heated space at the level desired and established by the thermostat setting. The line-voltage thermostat makes and breaks the actual operating current flowing to the heater. A low-voltage thermostat has its contacts arranged to make and break the low value of current to a low-voltage, 24-V operating coil of a relay, which in turn has its contacts arranged to make and break actual operating current to the heating load.

Each type of thermostat—line-voltage and low-voltage—has advantages and disadvantages. The low-voltage type, for example, can be made more sensitive to temperature changes because it only has to make and break the low current of the relay coil. Furthermore, where the line-voltage thermostat requires a switch box and standard power wiring for its installation, the low-voltage model can usually be simply screwed to the wall surface and connected with stapled low-voltage cable or bell wire for its low-voltage coil circuit. On the other hand, though, low-voltage controls are more expensive, and additional labor is involved in installing the relays. There are also more components to cause trouble in a low-voltage system.

Line-voltage thermostats (Fig. 5–3) are used separately to control heater units, but they are also built into heaters, using a sensing bulb in the heater airstream. This is probably the least expensive room control and is effective. However, such thermostats are slow-cycling and are not in the best position to sample temperature conditions in a room.

Although single-pole line-voltage thermostats are in common use where the *National Electrical Code*® permits, the double-pole line-voltage thermostats are normally considered to be the safest, mainly because they completely disconnect both "legs" of the 240-V power circuit, not just one leg as a single-pole thermostat does. With the two-pole type, the heater, while undergoing repairs, should then be completely free of any "live" wires.

One application of a special two-pole thermostat is in modulating control hookups for two-stage heaters. One set of contacts is designed to close ½ to 1½ degrees below the other set. Under moderate conditions, the first set of contacts close to bring on the first stage of the heater. If that stage cannot match the heat loss of the space and deep temperature at the desired level, the falling temperature will actuate the second set of contacts to turn on the second stage of the heater to increase the heat input to the heated space.

Low-voltage thermostats (Fig. 5–4) in combination with relays offer a wide range of control arrangements to meet practically any need that might be encountered. Two-stage low-voltage thermostats can be used with two re-

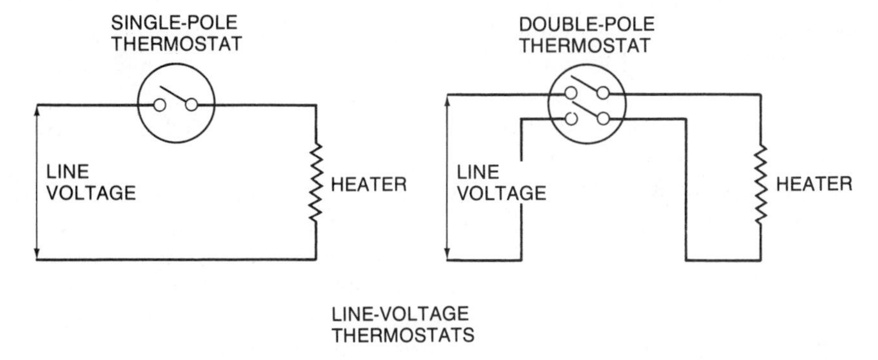

Figure 5–3. Two types of line-voltage thermostats.

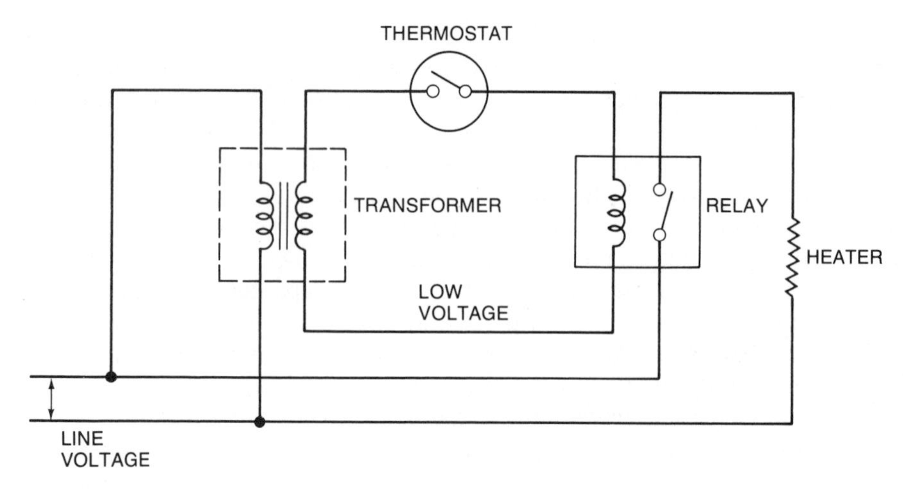

Figure 5–4. Typical low-voltage heating control circuit.

lays to connect a given heating unit from phase to neutral or from phase to phase on 120/240-V one-phase, three-wire circuit. In the 120-V connection, the heater output is only one-fourth that of its output when connected for 240-V operation. Again, this is done by having ½ to 1 degree of difference between the operating points of the two sets of contacts in the low-voltage thermostat.

Another control feature that may be used with electric heating systems is "night setback." Under most conditions it can be advantageous to lower the house temperature during sleeping hours and to bring the thermostat up to a high value about an hour before the family arises, so that the house will be relatively warm when they get out of bed. This is accomplished by means of a double thermostat working in conjunction with an electric timer. The controls, which can be set for any desired "night" and "day" periods, are arranged so that the two thermostats are connected in series. The day unit is always in the circuit, but during the day hours the night thermostat is merely short circuited. The timer opens this short circuit and puts the night control into action at night, and then takes it out of the circuit again in the morning. During the night, the furnace or heating units are controlled by the lower setting.

Controls are available which can also automatically sense nightfall. In any case, however, the control circuit of the thermostats is rearranged during night hours to set the thermostat to keep the temperature in the heated space at, say, 55°F or some other low level instead of the 70°F, which is the normal temperature to maintain during the daytime. This provides economy and meets the preference for lower temperature during sleeping hours.

TROUBLESHOOTING

In troubleshooting electrical heating units, it is often necessary to measure various electrical values, such as current, resistance, voltage, and so on. With an inexpensive volt-ohm-ammeter and a basic knowledge of its use, practically anyone can quickly determine the cause of most electrical problems that develop with the heating system. Then, with this knowledge, the person can decide whether repair of the problem is within his or her capabilities or whether a repairperson should be called.

The volt-ohm-ammeter is often referred to as a *multimeter.* These meters are available from electrical supply houses and electronic shops (such as Radio Shack) and sell anywhere from $25 to several hundred dollars depending on the quality. All of them will give good service if given the care they deserve. By the same token, if mistreated, any of them may fail to function properly.

When using any instrument for testing or measuring electrical circuits, always consider your personal safety first. Know the voltage levels and shock hazards related to all wiring and equipment to be tested. Also be certain that the instrument used for the application has been tested and calibrated properly.

When taking measurements with meters available with different ranges and functions, as in the case of a combination volt-ohm-ammeter, make certain that the meter selector and range switches are in the correct position for the circuit to be tested. For example, if voltage is being measured, make certain that the selector switch is set at VOLTAGE and is at the level of voltage (100, 150, 200, etc.) expected to be encountered. If there is any doubt as to what the exact voltage is, start at the highest voltage setting on the meter and work down as the pointer dictates.

As mentioned previously, the volt-ohm-ammeter is a great aid in troubleshooting electric heating systems. Troubleshooting covers a wide range of problems, from a small job such as finding a short circuit in, say, a baseboard heater to tracing out troubles in a complex control circuit. In any case, troubleshooting usually requires a basic knowledge of the testing instrument, and then going about the location of the problem in a systematic and methodical manner, testing each part of the circuit or system at a time until the trouble is located.

In general, there are only three basic electrical faults: a short circuit, an open circuit, or a change in electrical value.

Short Circuit: A short circuit is probably the most common cause of electrical problems. Basically, it is an undesired current path that allows the electrical current to bypass the load on the circuit. Sometimes the short is between two wires due to faulty insulation or it can occur between a wire and a grounded object such as the metal frame on an electric heater.

When a short circuit is suspected, disconnect all loads from the circuit and then reset the circuit breaker or replace the fuse. If this corrects the problem, it indicates that one of the heating units or devices is at fault. With the heaters still disconnected, connect the test leads of the ohmmeter to the heater leads, one test lead to each heater lead. If there is a full-scale reading on the meter, the short circuit is more than likely in that particular unit. You may, however, get only a partial reading on the ohmmeter scale; this is probably due to the resistance of the heating element. Try another unit until one is found that gives a full-scale reading.

Should the fault be in the circuit wiring itself (indicated by a tripped circuit breaker or blown fuse when all loads are disconnected from the circuit), the next step is to go along the circuit and open up the various outlet and junction boxes and so on until the trouble is located. In most cases, the short circuit will be found at one of the heater junction boxes where a terminal has vibrated loose. Perhaps one of the splices has become loose and is rubbing against the grounded heater housing.

At times, however, if repair or remodeling work has recently been done, the trouble may be caused by someone having driven a nail into a piece of nonmetallic cable, surface molding, or even through a piece of thin-wall (EMT) tubing; or the conductors may have accidentally been cut in two with a saw or drill.

Open Circuit: An open circuit is an incomplete current path like the broken wires in a circuit. Therefore, if the circuit is supplying an electric wall heater and the circuit is open, the heater will not operate. A light switch or thermostat, for example, purposely opens a circuit supplying a load. If the switch is then turned to the ON position and the light does not burn or the heater does not heat, the first assumption is that perhaps the lamp is bad or else the heating element has burned out. Both can be checked with the ohmmeter to determine if the element is good or bad. If either is good, the ohmmeter should show some resistance reading on the meter. If bad, there will be no reading on the meter. Should the element prove to be good, check for other problems in the circuit in the following order.

1. A blown fuse or a tripped circuit breaker
2. A wire loose at the switch or thermostat or in the fuse box
3. A faulty switch or thermostat

If the circuit in question is protected by a plug fuse, the nature of the problem can often be determined by the appearance of the fuse window. For example, if the window is clear and the metal strip appears to be intact, the fuse is probably not blown. But it is always best to check the fuse with a voltmeter or test lamp just to make sure. To test a plug fuse, place one lead of the voltmeter on the neutral block in the panelboard or fuse cabinet, and the other on the load side of the fuse. If a reading is obtained from between 110 and 120 V, the fuse is all right. If the voltmeter does not show any reading or one that is below 100 V, examine the fuse window more closely. If the window is clear but you notice that the metal strip is in two pieces, it was probably blown due to a light overload. Perhaps a portable appliance was plugged in on the same circuit. In any event, check to see what overloaded the fuse before replacing it.

Change in Electrical Value: A change in either current, voltage, or resistance from the normal can lead to electrical problems. One of the most common causes of electrical problems is low equipment input voltage. This problem usually occurs for one or more of the following reasons:

1. Undersized conductors
2. Loose connections
3. Overloaded circuit
4. Taps set too low on power company's transformer

To check for low voltage, set the multimeter to the voltage setting of 250 V and take a reading at the main switch or service entrance. In most cases, residential electric services are single-phase, 120/240-V, three-wire. Therefore, the voltage reading taken between any two "hot" wires (usually colored black on one conductor and red on the other) should be around 240 V, and the reading between any of the "hot" wires and the neutral (white) wire should be around 120 V. If the reading is below these figures, the fault lies with the utility company supplying the power, and they should be notified to correct the problem. However, if the reading checks out normal, the next procedure is to check the voltage reading at various outlets throughout the heating system.

When a low-voltage problem is found on a circuit—say, 100 V instead of the normal 120 V—leave the voltmeter terminals connected across the line, and begin disconnecting all loads, one at a time. If the problem is corrected after several of the loads have been disconnected, the circuit is probably overloaded and steps should be taken to reduce the load on the circuits or the wire size should be increased in size to accommodate the load.

Loose connections can also cause low voltage. The entire circuit should be deenergized, and each terminal of panelboard circuits, heater connections, and so on, should be checked for loose connections. A charred or blackened terminal screw is one sign that indicates this problem.

BASIC ELECTRICITY

Anyone anticipating troubleshooting electrical heating systems or the electrical/electronic controls in any type of heating system should have a good basic knowledge of electricity. The following is a brief review of the principles of electricity as applied to heating applications.

Flip a switch—a light comes on. Push a lever—bread is toasted. A live football game, being played hundreds of miles away, comes into focus at the touch of another switch. Put in the plug, pull the trigger switch, and a whole assortment of electric power tools perform immediately. Electricity is handy, convenient, taken for granted, and often not fully understood. It is well that you understand the fundamentals of electricity if any type of electric servicing will be performed while working on residential heating systems.

UNDERSTANDING ELECTRIC ENERGY AND HOW IT IS MEASURED

Some knowledge of electric energy is helpful. Such information as what it is, terms used, and how it is measured are described under the following headings:

- Understanding electric energy
- How electric energy is measured

Understanding Electric Energy: Electric energy is easier to describe by what it does rather than by what it is. Electricity is the moving of electrons from one atom to another (Fig. 5-5). The number of electrons in motion determines the amount of electric energy produced.

Electricity in its usable form is described under the following headings:

- Types of electric circuits
- Types of electric current

Types of Electric Circuits. When a wire is moved, it cuts the magnetic lines of force and electrical potential (voltage) is generated (Fig. 5-6). An *electric circuit* is a completed path for the exchange of electrons. The circuit causes current to flow from the source of power to its use and return to the source (Fig. 5-7). Circuits are usually made of wires called *conductors* (Fig. 5-8). They are protected by *insulation* (*nonconductors*) (Fig. 5-9).

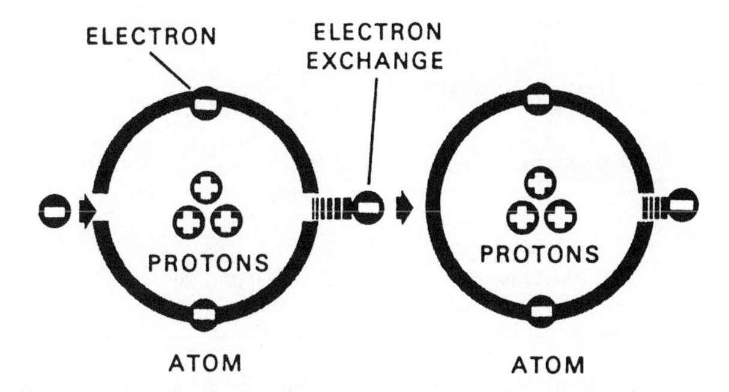

Figure 5–5. Electricity is the moving of electrons from one atom to another. (*Courtesy of American Association for Vocational Instructional Materials*)

Electric circuits are identified as follows:

- Series circuit (Fig. 5–10)
- Parallel circuit (Fig. 5–11)
- Open circuit (Fig. 5–12)
- Short circuit (Fig. 5–13)

Types of Electric Current. There are two types of electric current:

- Direct current
- Alternating current

Figure 5–6. Moving a wire through magnetic lines of force generates an electric potential in the wire. (*Courtesy of American Association for Vocational Instructional Materials*)

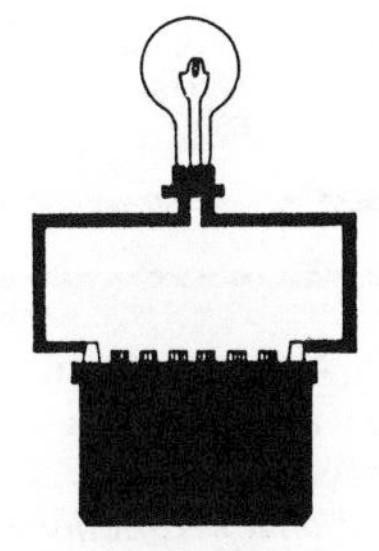

Figure 5–7. Current flows in a completed electric circuit. (*Courtesy of American Association for Vocational Instructional Materials*)

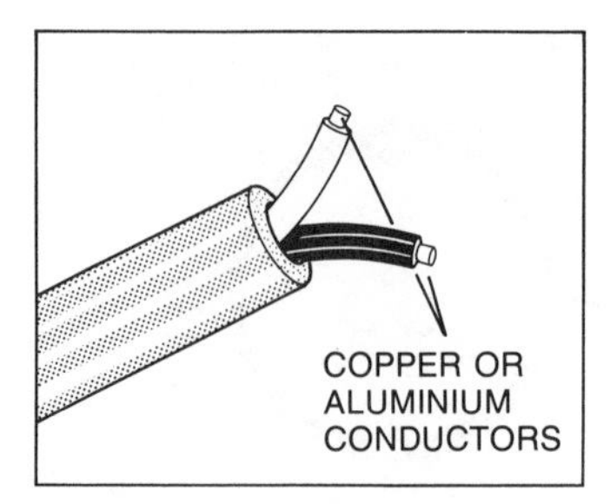

Figure 5–8. A conductor is a material that allows electron exchange. (*Courtesy of American Association for Vocational Instructional Materials*)

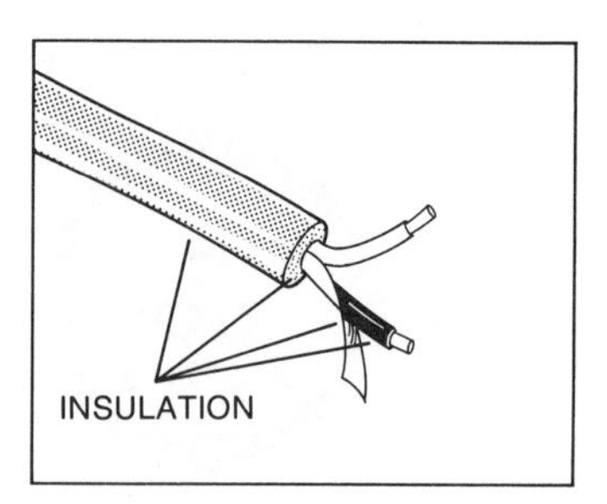

Figure 5–9. Insulation used with electric conductors is made of a material in which electrons are not exchanged. (*Courtesy of American Association for Vocational Instructional Materials*)

Figure 5–10. A series circuit is one in which all of the current flows throughout the entire circuit. (*Courtesy of American Association for Vocational Instructional Materials*)

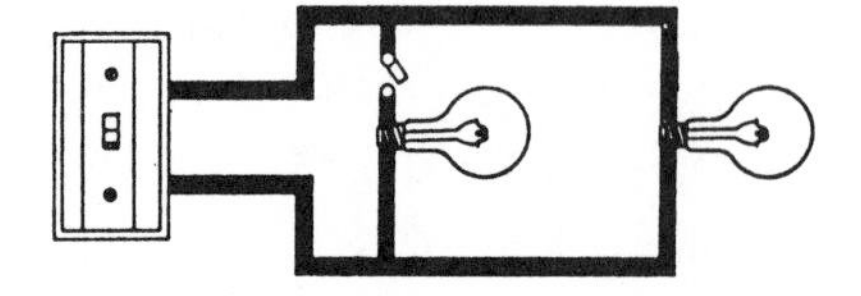

Figure 5–11. A parallel circuit provides for dividing the current flow. (*Courtesy of American Association for Vocational Instructional Materials*)

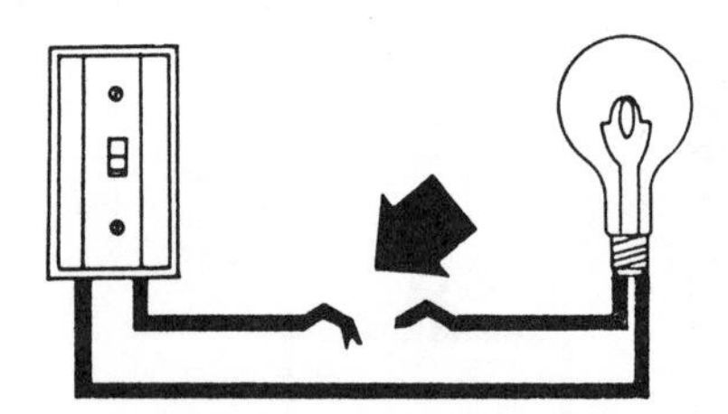

Figure 5–12. An open circuit is one that is broken and prevents current flow. (*Courtesy of American Association for Vocational Instructional Materials*)

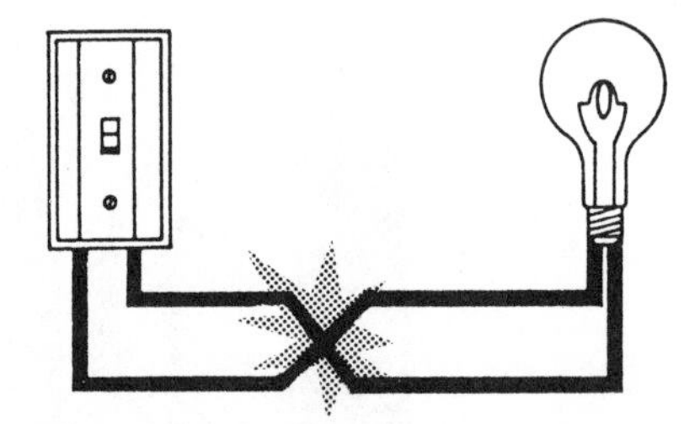

Figure 5–13. A short circuit is one that allows current to return to the source without traveling throughout the entire circuit. It may present an unsafe condition. (*Courtesy of American Association for Vocational Instructional Materials*)

Direct Current. Current flowing in only one direction in a circuit is called direct current (dc) (Fig. 5–14).

Alternating Current. Electrons flowing first in one direction and then in another are alternating current (ac) (Fig. 5–15). Alternating current is the kind of electric energy generally made available to you by electric power suppliers.

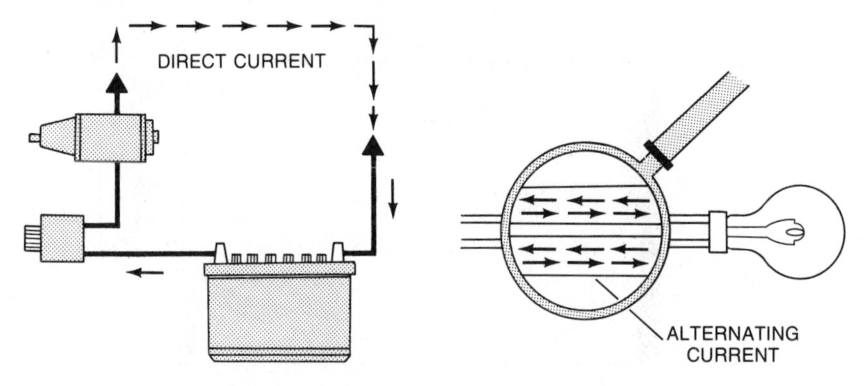

Figure 5–14. Direct current flows in one direction only. *(Courtesy of American Association for Vocational Instructional Materials)*

Figure 5–15. Alternating current flows in one direction and then in the other. *(Courtesy of American Association for Vocational Instructional Materials)*

Two special characteristics of alternating current are as follows:

1. Cycles per second (hertz)
2. Phases

1. CYCLES PER SECOND (HERTZ): A cycle is the flowing of alternating current in one direction, reversing, flowing in the other direction, reversing, and starting all over again.

The number of cycles per second is the frequency of the current. The unit for frequency is hertz (Hz). The frequency generated and distributed in most countries is 60 Hz (cycles per second). However, some countries use 50-Hz current.

2. PHASES: Electric energy is produced in phases: single-phase, two-phase, and three-phase. Single-phase current is the flowing of alternating current (Fig. 5–16). Two-phase current is sometimes produced, but has no advantages and is not generally used. Power plant generators are designed to produce three single-phase currents, each starting and reversing one-third of a cycle apart. They are combined to produce three-phase current (Fig. 5–17). Better conductor efficiency is obtained through three-phase current than through single-phase current. Electric motors operating on three-phase current are easier to start than those operating on single-phase current. Also, they are less expensive to purchase and maintain.

How Electric Energy Is Measured: Terms have been accepted by which electric energy can be measured. The following are some common terms:

- Ampere
- Volt

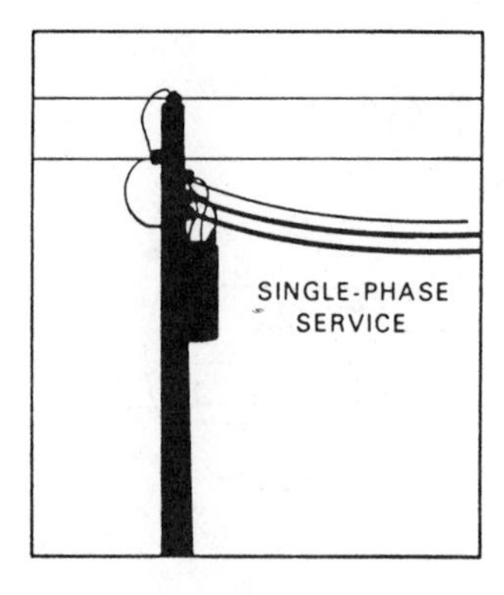

Figure 5-16. Single-phase service usually consists of three conductors between the power pole and the meter. (*Courtesy of American Association for Vocational Instructional Materials*)

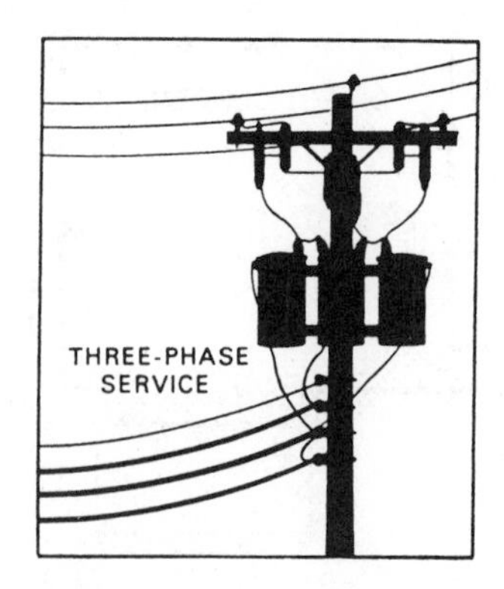

Figure 5-17. Three-phase service generally has four wires from the power pole to the meter consisting of three "hot" wires and a neutral wire. (*Courtesy of American Association for Vocational Instructional Materials*)

- Ohm
- Watt
- Kilowatt
- Megawatt
- Kilowatt-hour

An *ampere* is a unit of electric current. For example, a 100-W lamp bulb requires about 1 A of current on a 120-V system. Heat caused by friction of the electrons flowing through the lamp filament causes the lamp to glow. The instrument for measuring amperes (*amperage*) is called an ammeter (Fig. 5-18).

A unit of electric "pressure" that pushes the electrons along is a *volt;* the amount of pressure (generally referred to as *voltage*) required to push 1 ampere through 1 ohm of resistance is a volt. It is measured with a voltmeter (Fig. 5-19). The *ohm* is a unit of resistance to current flow. Electrons do not flow freely through any conductor. There is a certain amount of resistance overcome by 1 volt to cause 1 ampere to flow. It is measured with an ohmmeter (Fig. 5-20). Factors determining the amount of resistance in a conductor are material, size, and length (Fig. 5-21); temperature is another factor. A change in the temperature of a material changes the ease with which a conductor releases its outer electrons. For most materials the resistance increases as the temperature increases.

A *watt* (*wattage*) is a unit of electric power. It is measured with a watt-

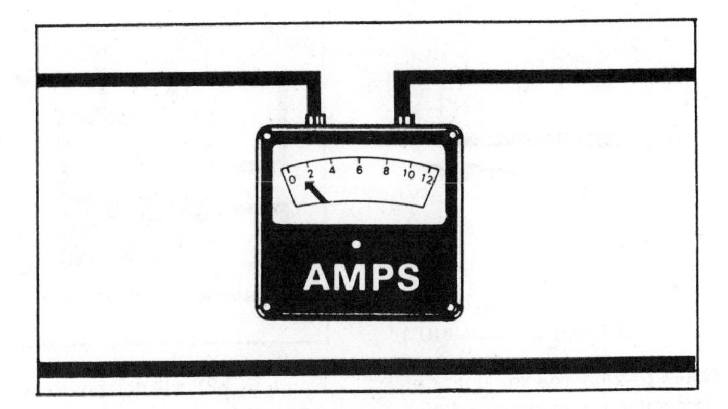

Figure 5–18. Electric current is measured in amperes with an ammeter. (*Courtesy of American Association for Vocational Instructional Materials*)

meter (Fig. 5-22). The combination of volts (pressure) and amperes (current flow) makes watts (power). Therefore,

$$\text{volts} \times \text{amperes} = \text{watts}$$

Example

120 volts × 10 amperes – 1200 watts. A *kilowatt* is 1000 watts. A *megawatt* is 1,000,000 watts. A *kilowatt-hour* is the unit of measure by which electric power is marketed. Therefore,

$$\text{kilowatts} \times \text{hours} = \text{kilowatt-hours}$$

If you operate a 1-horsepower motor (approximately 1 kilowatt) for 2 hours, you will use about 2 kilowatt-hours of electric energy.

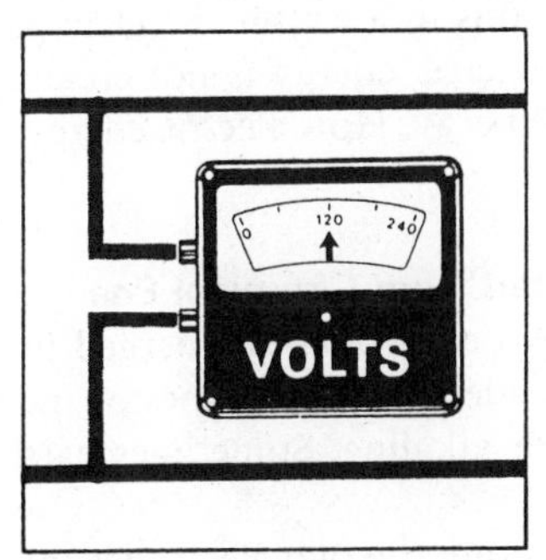

Figure 5–19. Electric voltage is measured with a voltmeter. (*Courtesy of American Association for Vocational Instructional Materials*)

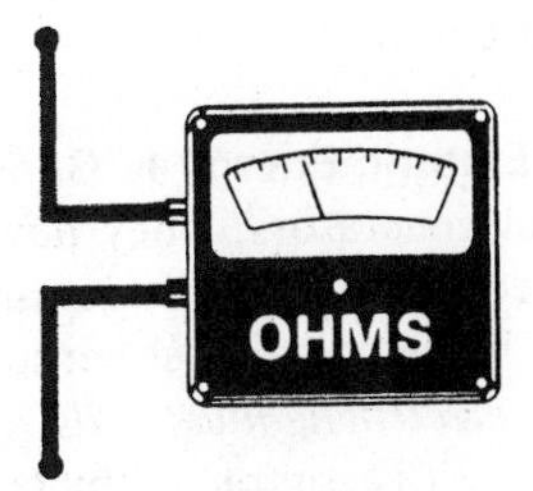

Figure 5–20. Resistance is measured in ohms with an ohmmeter. (*Courtesy of American Association for Vocational Instructional Materials*)

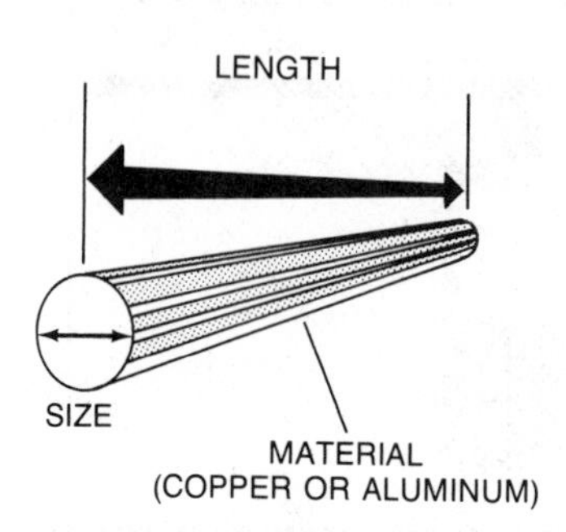

Figure 5–21. The amount of resistance in a conductor is determined by the type of material, size, and length. (*Courtesy of American Association for Vocational Instructional Materials*)

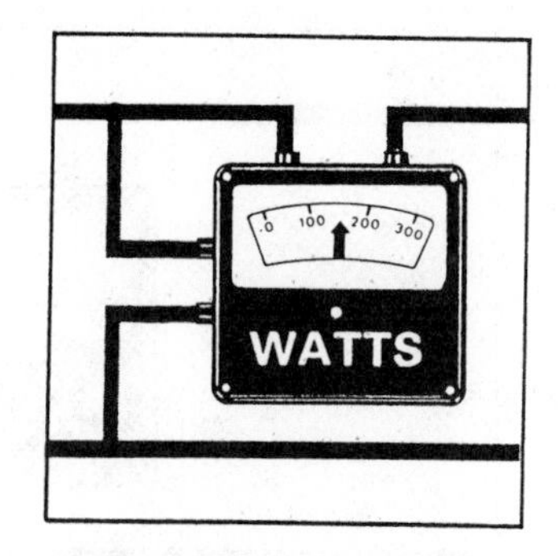

Figure 5–22. Electric power is measured in watts with a wattmeter. (*Courtesy of American Association for Vocational Instructional Materials*)

Example

$$1 \text{ kilowatt} \times 2 \text{ hours} = 2 \text{ kilowatt-hours}$$

The meter on your electric power entrance is a kilowatt-hour meter (Fig. 5-23).

UNDERSTANDING HOW ELECTRIC ENERGY IS GENERATED

Electric current is the flowing of electrons in a circuit. Electric power is generated by causing electrons to flow. Electric energy is not created; it is converted (generated) from other forms of energy. How electric energy is generated is described below.

How Electric Energy Is Generated from Chemical Energy: *Batteries* are chemical generators. They have two electrodes submerged in a chemical solution (electrolyte). Electrodes are made of various types of metal or other materials. The chemical is either acid or alkaline. Sometimes batteries are referred to as *electrochemical cells.*

The circuit is completed by connecting the two electrodes, at which time current flows within the circuit (Fig. 5–24). This flow is due to a chemical reaction of the electrolyte on the electrodes.

Batteries are made of one or more cells. Types of cells are as follows:

- Primary cells (Fig. 5–25)
- Storage cells (Fig. 5–26)

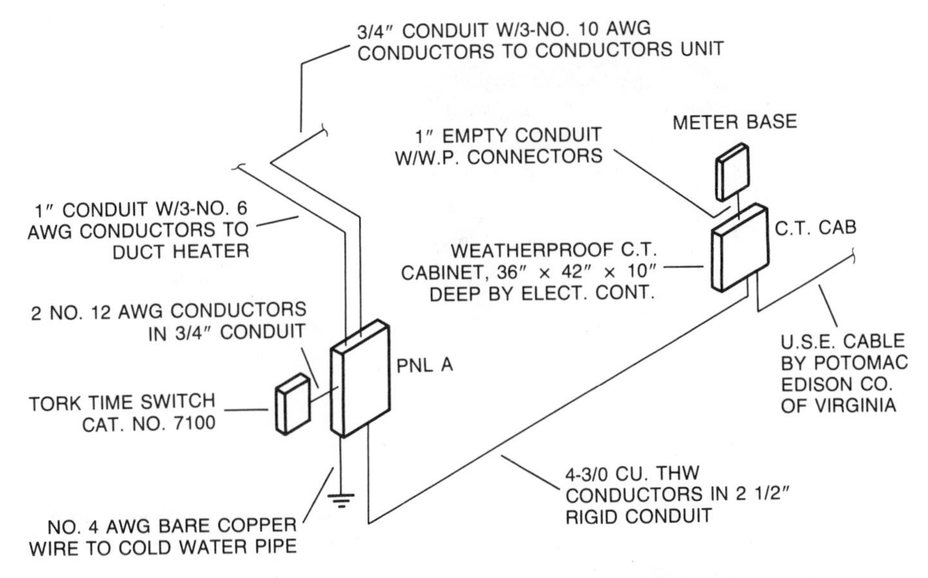

Figure 5–23. Electric energy used is measured by a kilowatt-hour meter. *(Courtesy of American Association for Vocational Instructional Materials)*

- Solar cells
- Fuel cells

How Electric Energy Is Generated from Mechanical Energy: Most electric energy is generated by mechanically driven generators. They are discussed under the following headings:

Figure 5–24. Current flows in a battery when the circuit is completed. *(Courtesy of American Association for Vocational Instructional Materials)*

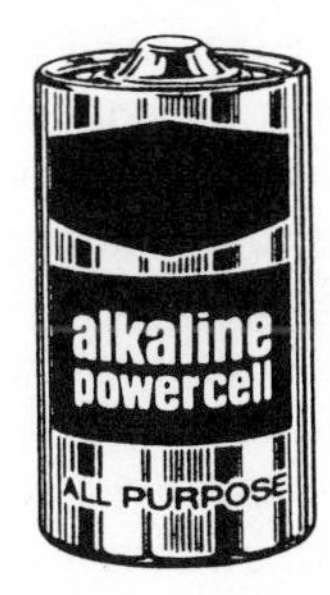

Figure 5–25. Primary dry cell battery. *(Courtesy of American Association for Vocational Instructional Materials)*

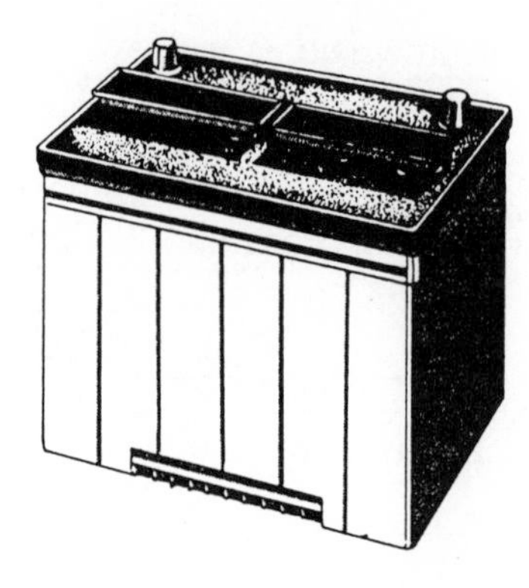

Figure 5–26. Automotive-type battery consisting of storage cells. (*Courtesy of American Association for Vocational Instructional Materials*)

- How mechanical generators work
- How mechanical generators are driven

How Mechanical Generators Work. Passing coils of wire through a magnetic field generates electric energy mechanically. Both direct and alternating current may be generated by mechanical generators. How this is done is discussed as follows:

- How direct-current generators work
- How alternating-current generators work

How Direct-Current Generators Work. The principles of the direct-current, mechanically driven generator are illustrated in Fig. 5–27. As magnetic "lines of force" travel from one pole to another, a simple horseshoe magnet forms a magnetic field. A loop of wire supported on a shaft is rotated within this magnetic field. An electric current is generated in this loop, as it turns, and the current flows to terminals (segments) on the commutator. Two stationary contacts, or brushes, touch these commutator segments as the loop of wire rotates. These brushes carry current generated in the loop to an external circuit, where it is used as electric energy.

Note that the commutator is in two halves. The loop of wire rotates one-half turn as each brush makes contact with the other half of the commutator. This is necessary to keep the current flowing in the same direction in the circuit; thus direct current is generated.

Direct current is used principally in mobile equipment and special industrial applications (Fig. 5–28). A serious disadvantage of dc current is that it is difficult to deliver over long distances.

How Alternating-Current Generators Work. An alternating-current generator, frequently called an *alternator,* operates on much the same principle as the dc generator (Fig. 5–29). A wire loop also rotates within a magnetic field; this rotation of the loop through magnetic lines of force generates current in the loop.

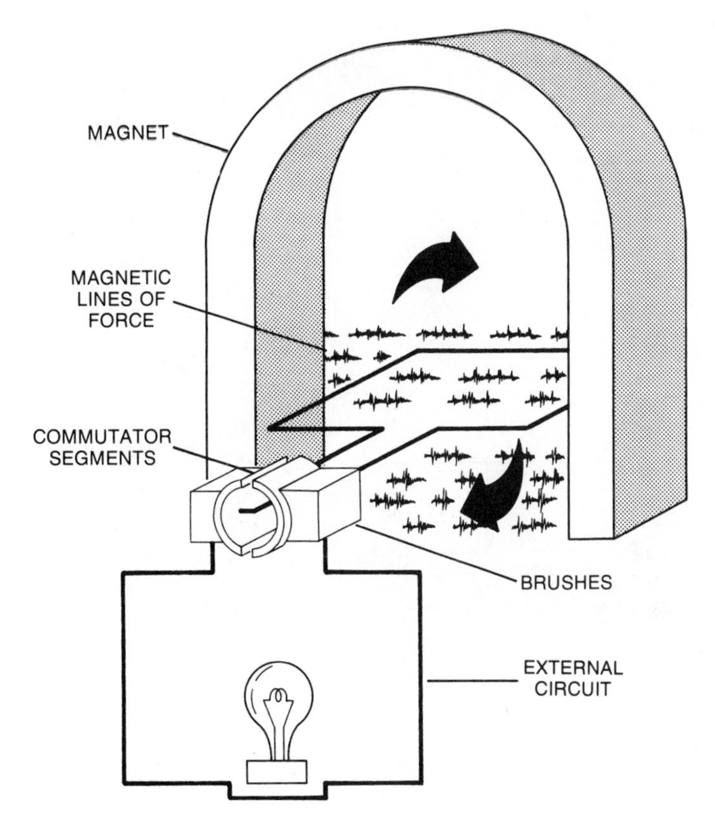

Figure 5–27. Principle of a simple direct-current generator. (*Courtesy of American Association for Vocational Instructional Materials*)

Instead of being connected to segments on a commutator, the loop ends are each connected to an individual ring called a slip ring. The slip rings are fastened to the shaft and rotate with it. Stationary brushes are in contact with the slip rings. Electric energy is conducted to its point of use by the current carried through the brushes to an external circuit.

During one revolution of the loop, each side passes through the magnetic lines of force: first in one direction and then in the other, thus producing alternating current.

How Mechanical Generators Are Driven. Mechanical generators are driven by water power and by internal combustion engines. These methods are explained as follows:

- Water power
- Steam power
- Internal combustion engines

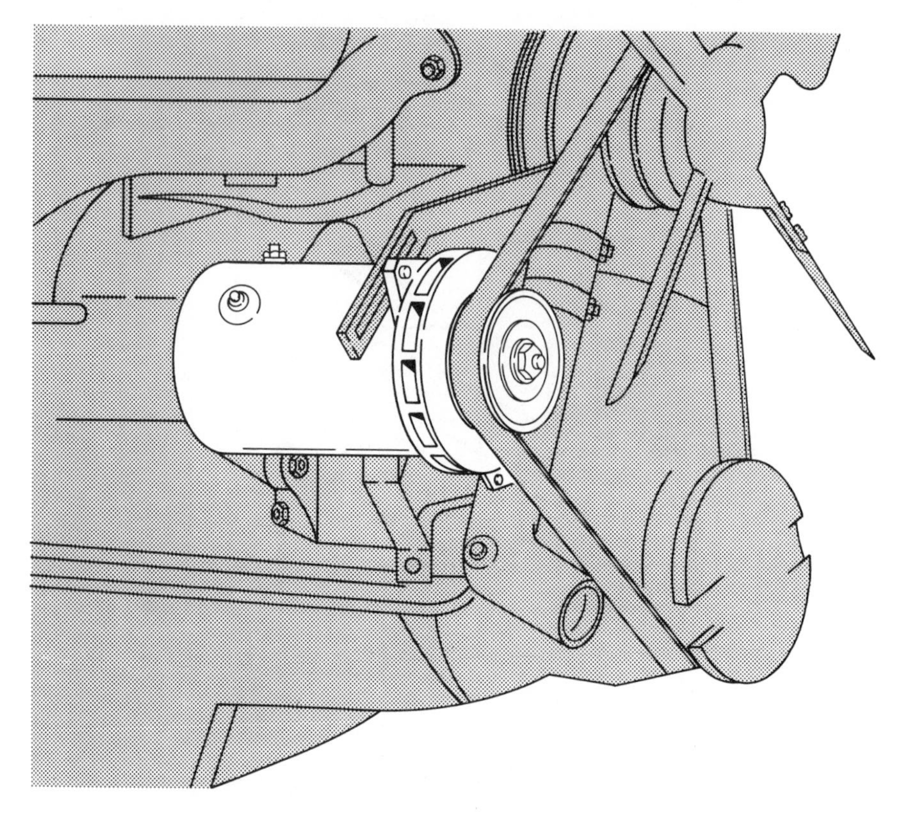

Figure 5–28. Automotive-type, direct-current generator. (*Courtesy of American Association for Vocational Instructional Materials*)

Water Power. Hydroelectric generation is water flowing from a higher to a lower level and being used to turn a generator. This type of electric energy generation developed very rapidly after 1910. By now, the best hydroelectric generation sites in North America have been developed. However, there are still many good sites remaining in the rest of the world.

Reservoirs are created by damming large streams, thereby providing a controlled flow of water. There may be more than one dam on the same stream. Water power uses the principle of the waterwheel to turn a generator. This "waterwheel" is made of metal and is called a *turbine wheel.* The generator is attached to the turbine shaft. Water directed against the turbine blades causes both the turbine and the generator to turn (Fig. 5–30).

Steam Power. Steam-powered generators produce about 85% of all electricity. Sometimes they are referred to as thermal-powered generators. The water is heated in a boiler until it becomes steam. At high temperatures and pressures the steam is directed against the blades or fins of the turbine, causing

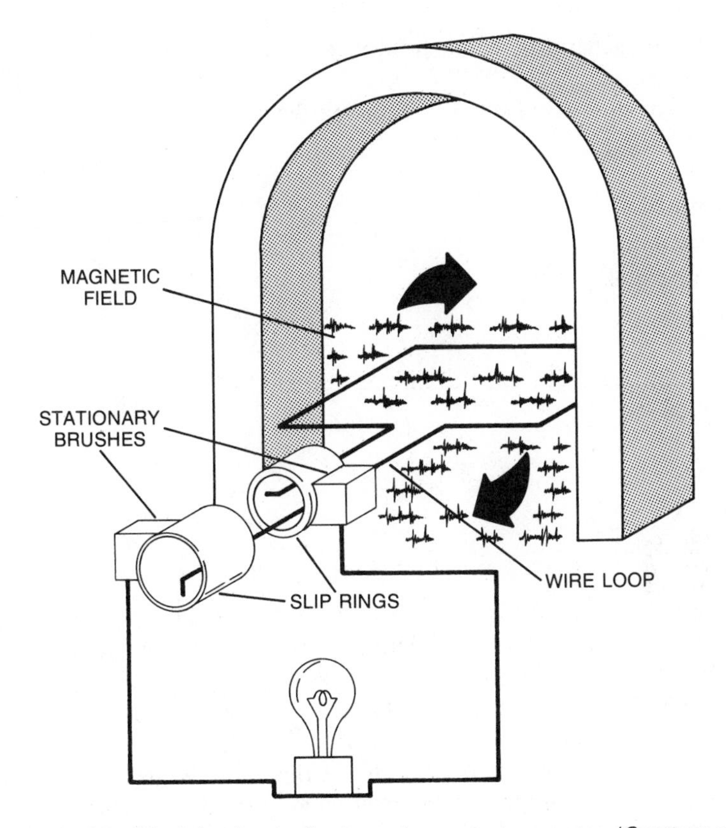

Figure 5–29. Principle of a simple alternating-current generator. (*Courtesy of American Association for Vocational Instructional Materials*)

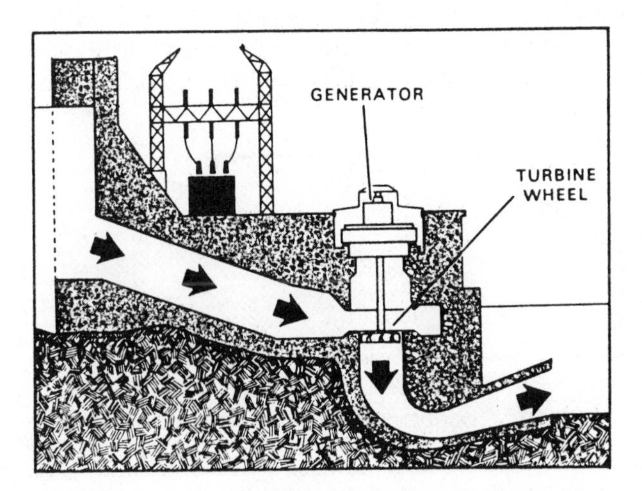

Figure 5–30. Principle of a water-powered generator. (*Courtesy of American Association for Vocational Instructional Materials*)

it to turn. The generator, being connected to the turbine, also turns (Fig. 5-31).

Heat for producing steam is supplied from the following sources of energy:

1. Fossil fuel
2. Nuclear
3. Geothermal
4. Solid waste

1. FOSSIL FUEL: The three kinds of fossil fuels used in producing steam are as follows:

- Coal
- Oil
- Natural gas

Coal is the fuel most commonly used to make steam for generating electric energy. It can be obtained at relatively low cost and is in abundant supply. It is found in many regions of the world. When available, a 90-day supply is usually maintained at most power plants. Conveyor belts carry coal from a coal pile into the plant. Massive pulverizers literally reduce the coal to dust. Coal dust is then blown into the firebox of the boiler under high pressure. It is burned almost instantly, creating heat which produces steam (Fig. 5-32).

Oil is also used for fuel in steam generating plants (Fig. 5-33). Heavy residual oil is used most commonly. Oil, being easier to handle and cleaner burning, was preferred over coal for many years. For that reason many plants were converted from coal to oil. However, some of these plants are being converted back to coal because of the increased cost of oil and the uncertainty of an adequate supply.

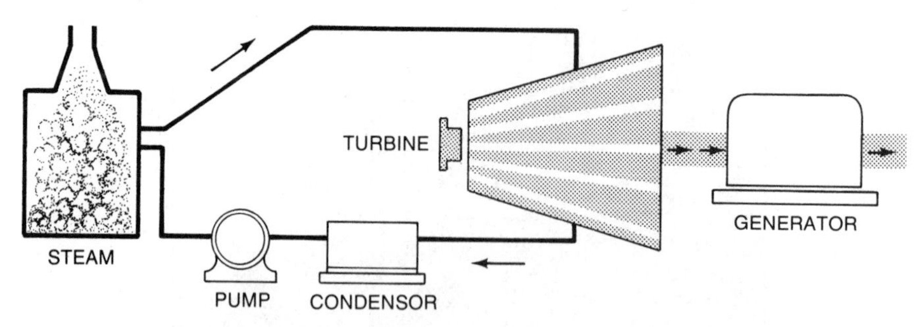

Figure 5-31. Principle of a steam-powered generator. (*Courtesy of American Association for Vocational Instructional Materials*)

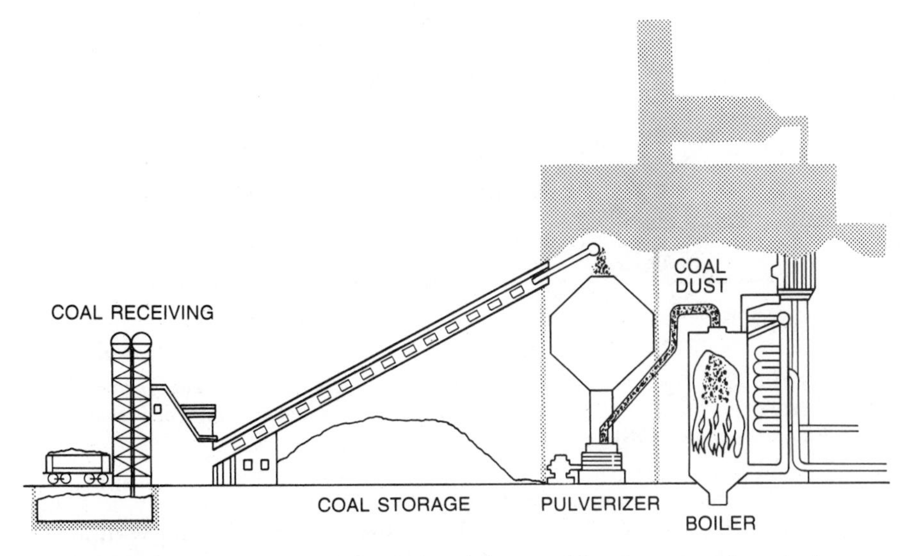

Figure 5–32. How steam is produced in a coal-burning generating plant. (*Courtesy of American Association for Vocational Instructional Materials*)

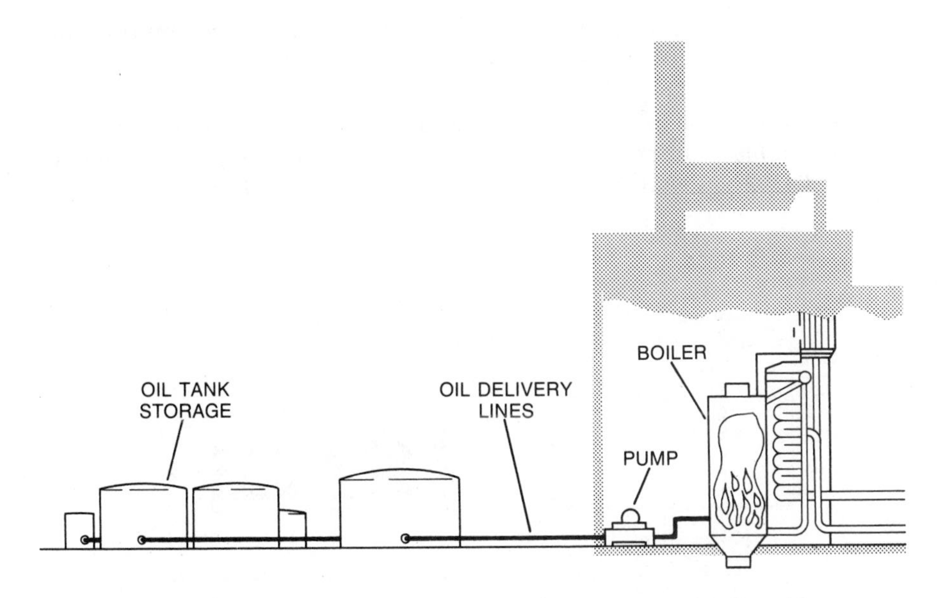

Figure 5–33. How steam is produced in an oil-burning generating plant. (*Courtesy of American Association for Vocational Instructional Materials*)

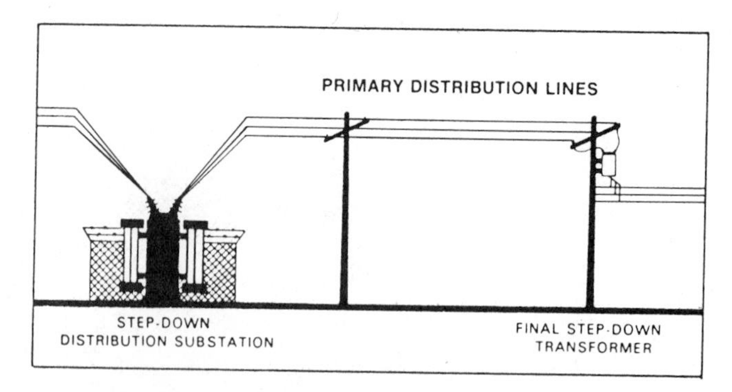

Figure 5–34. Primary distribution lines are lines carrying electric energy from a step-down transformer. (*Courtesy of American Association for Vocational Instructional Materials*)

Natural gas is used in some steam generating plants. Generally, it is supplied on what is called an "interruptible" basis, whereby natural gas suppliers can limit or stop delivery during such times as the winter season when additional gas is needed for heating. At that time electric power suppliers must switch to another fuel, usually oil.

Internal Combustion Engines. A limited number of electric generators are powered by internal combustion engines. These engines may be of the following types:

1. INTERNAL COMBUSTION TURBINE. Internal combustion turbines (heavy-duty jet engines) are used by power suppliers only when a high electrical demand is placed on the system. This limited use is because of the high cost of fuel and low operating efficiency to the unit. Power generated by this method may cost more than the amount charged by the power company.

2. DIESEL ENGINES. Standby generators operated by diesel engines

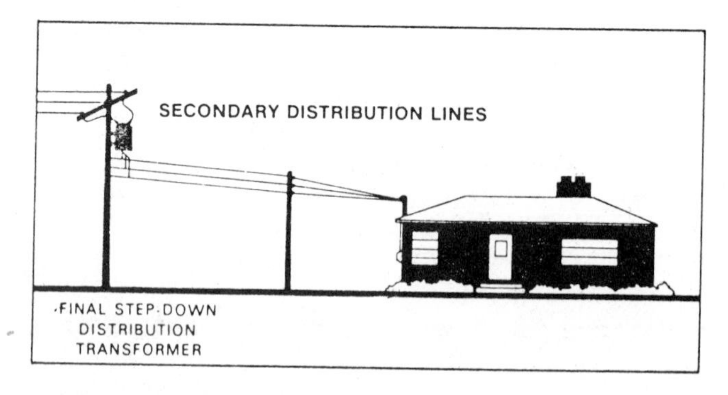

Figure 5–35. Secondary distribution lines are lines carrying electric energy from the final step-down transformer to the user. (*Courtesy of American Association for Vocational Instructional Materials*)

are used by such facilities as hospitals and manufacturing plants in case of power failure.

3. GASOLINE ENGINES. Portable generators or alternators are powered by both gasoline and diesel engines. They can be mounted on trucks, tractors, trailers, skids, or carrying frames for convenience.

USE OF DISTRIBUTION LINES

Lines that fan out over the countryside distribute electric energy from step-down distribution substations; thus power is delivered to the user. Generally, there are two types of distribution lines:

- Primary with voltages ranging from 2300 to 34,500 V. Typical primary line voltages are 2300, 4000, 8000, 12,000, 13,800, 20,000, 25,000, and 34,500 V (Fig. 5-34).
- Secondary with voltages usually of 120, 208, 240, 277, and 480 V (Fig. 5-35).

6

HEAT PUMPS

The term *heat pump,* as applied to a year-round air-conditioning system, commonly denotes a system in which refrigeration equipment is used in such a manner that heat is taken from a heat source and transferred to the conditioned space when heating is desired; heat is removed from the space and discharged to a heat sink when cooling and dehumidification are desired. Therefore, the heat pump is essentially a heat-transfer refrigeration device that puts the heat rejected by the refrigeration process to good use. A heat pump can

1. Provide either heating or cooling
2. Change from one to the other automatically as needed
3. Supply both simultaneously if so desired

A heat pump has the unique ability to furnish more energy than it consumes. This uniqueness is due to the fact that electric energy is required only to move the heat absorbed by the refrigerant; thus a heat pump attains a heating efficiency of 2 or more to 1; that is, it will put out an equivalent of 2 or 3 W of heat for every watt consumed. For this reason its use is highly desirable for the conservation of energy.

Compared to air-to-air heat pumps, water-to-air heat pumps, have the following advantages:

1. The pumps can be located anywhere in the building since no outside air is connected to them.

2. The outside air temperature does not affect the performance of the heat pump as it does an air-to-air type of heat pump.
3. Since the water source of the water-to-air heat pump will seldom vary more than a few degrees in temperature, a more consistent performance can be expected.

The heat pump is controlled automatically from a centrally located room thermostat measuring the temperature in the area and sending electrical control signals back to the heat pump system to control the heating or cooling operation.

Cooling Operation: In the summer, when an area needs cooling to make living conditions comfortable, the room thermostat will automatically send an electrical signal back to the heat pump, causing the unit to turn on in its cooling mode (Fig. 6–1). In the cooling mode, the compressor located outdoors in the outdoor section of the heat pump system will come on, together with the quiet air-moving fan on the top of the outdoor unit. The blower motor in the electric furnace will also come on to circulate the cool air. The heat pump system actually cools the area by removing heat from the air. As the warm air from the area is recirculated by the furnace blower, it passes over a V-coil which is at a very low temperature, thus removing heat. This performs a second important function, removing moisture from the air. As the warm air passes over the V-coil in the furnace, moisture in the air condenses on the coil much the same as moisture in the air condenses on the outside of a glass of ice water in a warm room. This moisture is drained off, leaving cool and comfortable air in the area.

Heating Operation: In the winter, when heating is needed, the room thermostat will send an electrical signal back to the heat pump, causing the unit to turn on its heating mode. In the heating mode, the compressor in the outdoor section of the heat pump system will come on, together with the air-moving fan in the outdoor section. The blower in the indoor section of the heat pump system, also known as the electric furnace, will come on to circulate warm heated air throughout the area.

It is normal for the unit to cycle on and off up to four or five times per hour on mild days. If the unit cycles more than four or five times each hour, a slight adjustment should be made to the thermostat heat anticipator.

Air heated by a heat pump will normally be discharged at a temperature of between 90 and 120°F, depending on the outdoor temperature. This air may not feel as warm as the air that discharges from a gas or oil furnace, but it contains the necessary amount of heat to raise the indoor temperature to a comfortable level and maintain it throughout the residence. Because of this design feature, a heat pump is capable of maintaining a balanced temperature rather than cycling with blasts of extremely hot air and then shutting off for longer periods of time, as is customary with other types of heat.

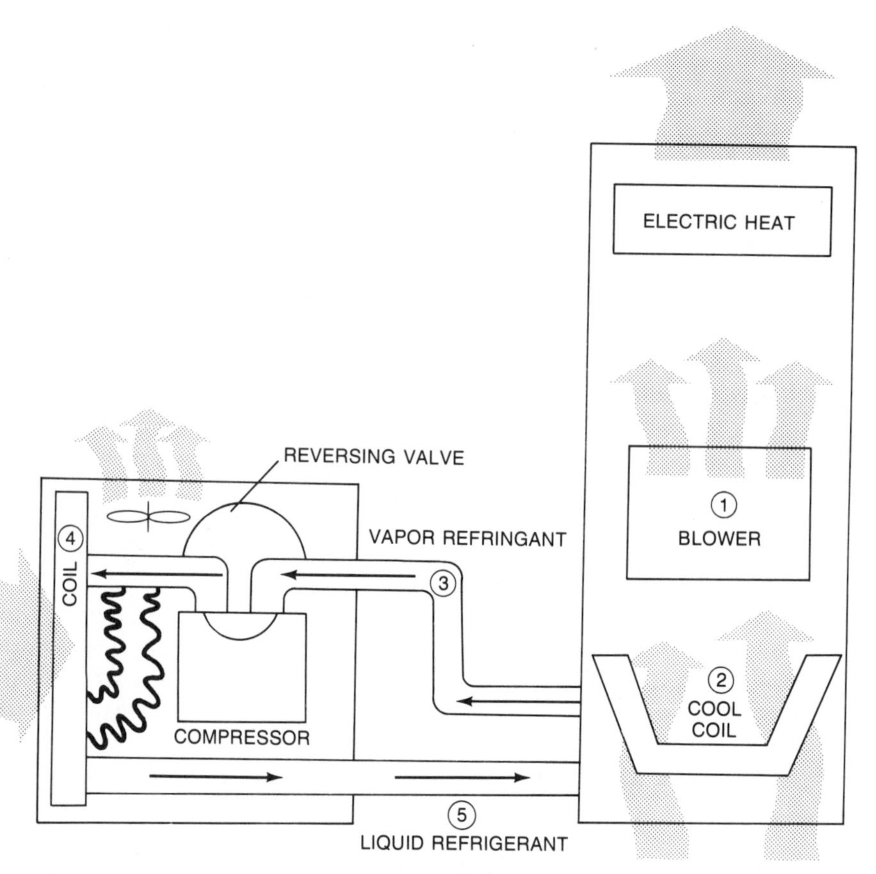

Figure 6–1. Basic operating principle of a heat pump in the cooling mode. For heat, the cycle is reversed. Should the outside air temperature reach too low a level for adequate heating, the auxiliary electric resistance heating coils energize for additional heat. (*Courtesy Square D Company*)

During the heating mode, a special device in the outdoor section, called a *reversing valve,* operates to control the direction of refrigerant flow. It will make a slight whoosing sound when it is first turned on. This is a normal operating sound and indicates that the system is functioning correctly. This whoosing sound can also be heard when the unit turns off or goes through a defrost cycle.

If the area in which the heat pump is located is in a part of the country where the outdoor temperature in winter drops below 40°F, the heat pump system will have to be equipped with auxiliary heating coils located either in the ductwork or else in the furnace itself. These coils will automatically come on to add extra heating capacity to your total heat pump system to heat the area on cold days. These boost heating coils are automatically controlled by the room thermostat. The heat pump will operate to provide efficient heat down to temperatures as low as −20°F. When the outdoor temperature goes below −20°F, some of the heat pump automatic internal safety controls will turn off the heat pump, allowing the area to be totally heated by the boost heating coils in the electric furnace. If the room thermostat is equipped with a check light, this check light will come on to indicate that the heat pump is not operating. The system will automatically reset after the outdoor temperature has come up above −20°F. This reset will take place during a thermostat cycle.

Defrost Cycle: When the outdoor temperature is below approximately 45°F, a small amount of frost will begin to form on the outside surface of the refrigerant coil on the outdoor unit. This is also a normal operating condition for a heat pump system. The unit will automatically defrost periodically (usually after every 90 minutes of running time) to remove this coating of frost and prevent it from building up on the coil surface. This defrost operation is controlled by automatic sensors in the unit which act to switch the reversing valve at the beginning and end of each defrost cycle. A normal defrost cycle will last between 1 and 6 minutes. During the defrost cycle, the outdoor fan motor will stop and the coil will heat up to melt away the frost. This melting results in the release of a quantity of steam from the outdoor unit. At the conclusion of a defrost cycle, the air-moving fan will again restart automatically, exhausting the remainder of the steam from inside the unit.

The location of the thermostat that controls a heat pump is most important to ensure that it will provide a comfortable temperature. In general, the thermostat should be located about 5 ft above the floor and installed on an inside wall. Never expose it to direct light from lamps, sun, fireplaces, and so on. Avoid windows, adjoining outside walls, or locations close to doors that lead outside. Locations close to radiators, warm-air registers, or in the direct path of heat from them should be avoided. Make sure that there are no pipes or ductwork in the part of the wall chosen. Never locate it in a room that is warmer or cooler than the rest of the home, such as the kitchen or hallway. The living or dining room is normally a good location, provided that there is no cooking range or refrigerator on the opposite side of the wall.

Efficiency: The heat pump is one of the most efficient heating systems known. It is capable of producing more heating energy than it consumes in electricity, resulting in efficiencies as high as 300%. It may sound impossible, when you realize that gas and oil furnaces normally have efficiencies of only 60% to 80%, and electric furnaces and baseboard heating equipment can only operate at 100% efficiency. The heat pump system operates at these seemingly impossible efficiencies because it not only generates heat itself, but transfers heat from the outside air (which has been heated by the sun) inside the residence together with its own heat. Even air at 0°F contains a certain amount of heat. By means of the refrigeration cycle, heat is extracted from the outside air and literally "pumped" inside the residence (hence the name "heat pump"). Thus the heat pump system can provide more energy in the form of heat to the residence than it consumes, helping to keep electric bills as low as possible. The features of heat pumps vary somewhat from manufacturer to manufacturer, but the following is typical of one manufacturer.

Electric Expansion Valve: All heat pump systems use an expansion device to control the flow of refrigerant. One system employs a unique refrigerant control device consisting of an electric expansion valve and two thermistors to monitor refrigerant flow and continuously modulate flow rates to achieve the highest possible efficiency at all operating conditions. Refrigerant engineers refer to this type of system as a "zero superheat" control. This means that all the refrigerant is constantly being used to heat or cool your home. In other types of systems, some of the refrigerant is pumped around the system without actually working to remove heat. This can be measured in terms of temperature and pressure, and is referred to as *superheat.* The electric expansion valve is the only control system that constantly senses liquid refrigerant directly and modulates to maintain zero superheat at the reversing valve.

Crankcase Heater: Most heat pumps are equipped with a compressor crankcase heater that keeps the compressor temperature warmer than surrounding air even on very cold days. This helps to prevent liquid refrigerant from migrating to the compressor, a common cause of compressor failure.

High-Pressure Control: All heat pumps are equipped with a safety high-pressure control that will automatically shut the unit off in the event of a fan motor failure or other blocked-coil condition. This will help protect the compressor from the tremendous stress that can occur when a fan motor fails or airflow is cut down for any reason. It is important for you to keep the filters in your system clean to help prevent the high-pressure control from operating and to keep your system at its highest operating efficiency.

Low-Pressure Control: A heat pump is also equipped with a low-pressure control to protect you from an expensive compressor failure if the unit loses its refrigerant charge for any reason. Sometimes a tube will be crushed or damaged by a lawn mower or punctured in some other fashion, and refrigerant can leak out of the system. If the unit were to operate without refrigerant, there is a high probability that the compressor would fail. To prevent this expensive failure, most heat pumps have a special low-pressure safety control to shut off the compressor automatically if there is a loss of charge condition.

Suction Line Filter/Dryer: A heat pump contains a special suction line filter/dryer that protects the compressor against any contamination in the refrigerant system. It is important to note that the filter/dryer is located in the suction line where it can provide the maximum protection to the compressor rather than in some other part of the system.

Time/Temperature Defrost System: The defrost controls built into the heat pump have been proved by many years of field testing. The time/temperature concept provides a high degree of reliability in this important area of heat pump control. This automatic control will sense the temperature of the outdoor coil and permit it to defrost every 90 minutes if a defrost cycle is required. This frequent defrosting will enable the unit always to operate at peak efficiency during a heating cycle, keeping your electric bill as low as possible.

Suction Line Accumulator: A special suction line accumulator built into the heat pump protects the compressor from excessive refrigerant floodback on cold operating days.

Troubleshooting heat pumps and related systems cover a wide range of electrical and mechanical problems, from finding a short circuit in the power supply line, through tracing loose connections in complex control circuits. However, by using a systematic approach, checking one part of the heat pump's system at a time, the cause of the trouble can usually be found.

CHECK TEST AND START

All heat pumps are equipped with a system crankcase heater. The purpose of this heater is to remove liquid refrigerant from the compressor crankcase and prevent it from accumulating in the compressor. On new systems it is extremely important that this crankcase heater be energized at least two hours prior to starting the system in order to protect the compressor. The crankcase heater will automatically operate whenever power is connected to the power terminals of the circuit breaker and the circuit breaker is in the ON position.

To check the cooling cycle, a low-pressure gauge should be connected to the suction line service port located on the compressor suction line adjacent to the compressor. A high-pressure gauge should be connected to the service port provided in the interconnecting tubing between the reversing valve and the compressor. The access cover should be replaced, and the gauge should be allowed to extend outside the unit between the access door and the valve panel.

To check the cooling cycle, set the indoor thermostat to the system cooling or auto position and lower the thermostat to its lowest setting. The system should start on the cooling cycle. Allow the system to operate for 15 minutes and then check the pressure readings. These readings should be approximately the same as those readings shown in Table 6-1. If the temperatures are lower than the reference temperatures in the tables, pressures will be slightly lower.

TABLE 6.1. COOLING CYCLE DATA BASED ON 400 CFM PER TON AND 80° DB-670WB AIR ON INDOOR COIL. TESTED WITH 25 FT OF INTERCONNECTING TUBING.

Outdoor Temp °F	Discharge Pressure PSI		Suction Pressure PSI	
	HS 2.5	HS 3.0	HS 2.5	HS 3.0
60	170	160	58	56
70	200	190	66	62
75	215	205	67	65
80	230	220	68	67
85	242	235	70	70
90	256	250	71	72
95	270	266	72	74
100	284	282	74	77
105	298	298	75	78
110	314	314	77	80
115	330	332	79	81

(Courtesy Square D Company)

It is recommended that the indoor unit be checked for noise and vibration and that the duct system be checked to ensure that there is proper air distribution. If any adjustment needs to be made to either the indoor or outdoor unit, power should be disconnected prior to removing the service access panels.

HEATING-CYCLE CHECK

The heating cycle can be checked by placing the thermostat in the heat or auto position and moving the temperature indicator to its highest setting. The system reversing valve will reverse and the system will operate in its heating cycle.

TABLE 6.2. HEATING CYCLE DATA BASED ON 400 CFM PER TON 70° AIR ON INDOOR COIL. TESTED WITH 25 FT. INTERCONNECTING TUBING.

Outdoor Temp °F	Discharge Pressure PSI		Suction Pressure PSI		Indoor Coil Air Off Temp °F
	HS 2.5	HS 3.0	HS 2.5	HS 3.0	HS 2.5 & HS 3.0
−10	155	140	15	10	80
0	158	150	17	14	82
10	165	164	22	21	84
20	182	180	29	29	88
30	204	198	36	38	93
40	228	216	43	47	96
50	252	236	50	56	100
60	274	255	61	66	104
70	298	276	72	81	108

(Courtesy Square D Company)

Then after 15 minutes of running time, recheck the system pressures. Readings should approximate the readings shown in Table 6–2.

If the system is equipped with supplemental heat it will be necessary to disconnect the appropriate leads on the electric furnace temporarily to prevent supplemental heat from coming on while taking air off temperature readings.

DEFROST-CYCLE CHECK

While running on the heating cycle, the defrost cycle can be checked by completely blocking the air inlet to the outdoor coil and the air discharge from the outdoor fan. It will be necessary to remove the access door to the control compartment. After 5 to 7 minutes of running time, rotate the screwdriver slot on the defrost control until it clicks audibly. This should happen within one full revolution.

The reversing valve should reverse and the outdoor fan motor should stop. The unit should return to normal operation in less than a minute if the outdoor temperature is above 60°F. The defrost cycle may take slightly longer if the temperature is lower.

On systems equipped with accessory "Auxiliary Heat Only" control and a room thermostat with an "Auxiliary Heat Only" position, a check may be made while the system is operating in the heating mode. Switch the small system switch on the room thermostat to the "Auxiliary Heat Only" position. The outdoor unit should stop and the indoor electric strip heat should come on. All heaters should sequence on. The system "Check" (usually a light on the thermostat) should come on. After completing the check, return the system switch to the appropriate COOL, HEAT, or AUTO position.

MAINTENANCE

Most of the higher-quality heat pumps will require very little maintenance. However, the outdoor unit fan motor should be lubricated annually with No. 20 grade nondetergent oil. Use about 40 to 50 drops in each oiler tube or as recommended by the manufacturer. Over-oiling can damage the motor, so use caution here and never use more than the amount recommended.

The outdoor coil should be cleaned periodically to remove accumulated dirt, leaves, and debris. Keeping the coil clean will result in the highest possible efficiency of operation.

REFRIGERANT PIPING

Excessive refrigerant charge is a major cause of compressor failure and the heat pump is especially susceptible to overcharging. Most models are fitted with an accumulator in the suction line for storage of excess refrigerant on the heating cycle and defrost cycle. However, care must be taken in charging to prevent overcharging, as the accumulator is sized only for the proper system charge. In systems with unavoidable long lines, additional charge will be required, but carefully follow the recommendations of the manufacturer when doing so.

For proper application, follow the procedure outlined below.

1. Determine the basic arrangement (A, B, or C) based on the relative elevation of the indoor and outdoor unit sections. See Fig. 6-2.

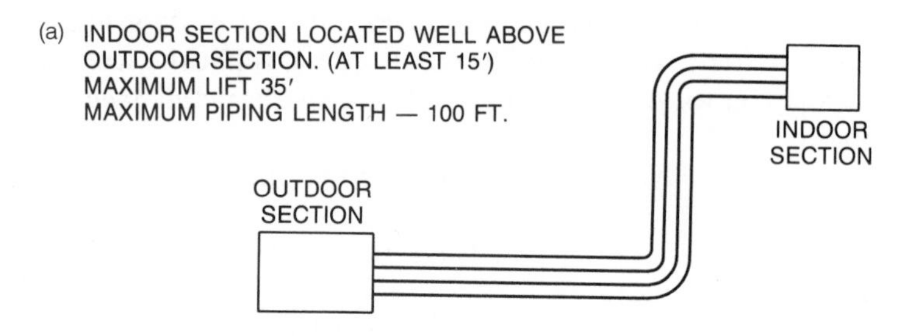

(b) INDOOR SECTION AND OUTDOOR SECTION ON OR ABOUT THE SAME LEVEL.
MAXIMUM PIPING LENGTH — 100 FT.

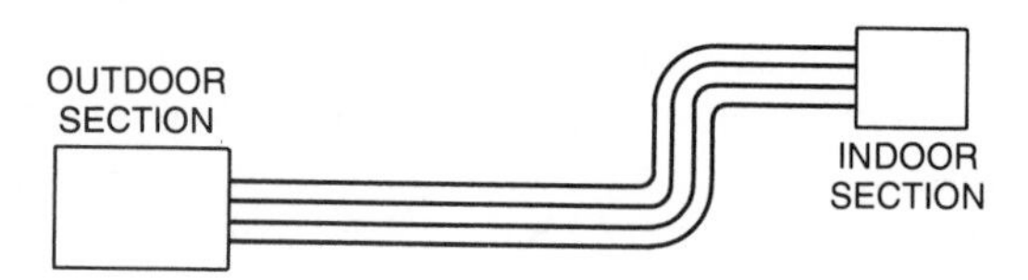

(c) MAXIMUM LIFT — 35 FT.
MAXIMUM PIPING LENGTH — 100 FT.
INDOOR SECTION LOCATED WELL BELOW OUTDOOR SECTION. (AT LEAST 15′)

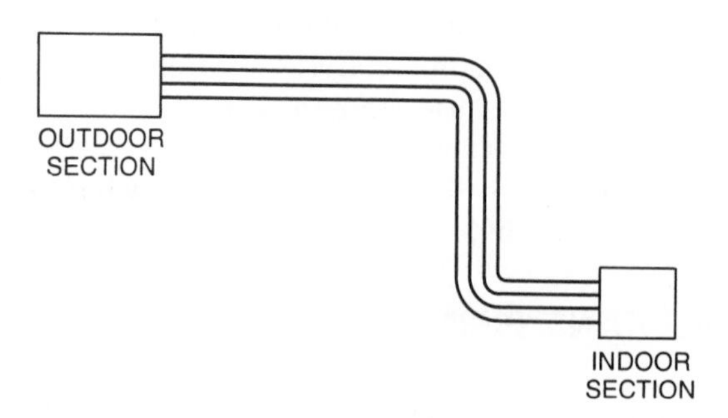

Figure 6-2. (*Courtesy Square D Company*)

2. Use Table 6–3 to determine the correct liquid line size. These sizes have been selected to provide the minimum pressure drop and to prevent liquid from flashing due to the pressure drop of long line lengths. The pressure drop in the vertical arrangement will be approximately 0.5 psig (pounds per square inch, gauge) per foot of height difference, in addition to the friction losses. Maximum lift should be limited to 35 ft, as severe capacity losses may result above 35 ft. Both refrigerant lines used to field pipe the heat pump must be insulated.

   Line lengths of over 100 ft are not recommended. If line lengths of over 100 ft are necessary, consult with the manufacturer and provide complete information concerning the application before making the installation.

   Care must be taken when installing field tubing to prevent collapse or kinking. Use a standard-type tube bender for proper bending where practical. If tubing is bent by hand, extra care must be exercised to prevent kinking, so do not remove the end plugs in the tube until connections are ready to be made.

TABLE 6.3. LIQUID LINE SIZES

Model	STD Liquid Line	Arrangement	Distance Between Indoor and Outdoor Section			
			Up to 25′	25′ to 50′	50′ to 75′	75′ to 100′
HS1.5	1/4	A	1/4	1/4	1/4	1/4
		B	1/4	1/4	1/4	1/4
		C	1/4	1/4	1/4	1/4
HS2.0	1/4	A	1/4	1/4	5/16	5/16
		B	1/4	1/4	5/16	5/16
		C	1/4	1/4	1/4	5/16
HS2.5	5/16	A	5/16	5/16	5/16	3/8
		B	5/16	5/16	5/16	3/8
		C	5/16	5/16	5/16	3/8
HS3.0	5/16	A	5/16	5/16	3/8	3/8
		B	5/16	5/16	3/8	3/8
		C	5/16	5/16	5/16	3/8
HS3.5	5/16	A	5/16	5/16	3/8	3/8
		B	5/16	5/16	3/8	3/8
		C	5/16	5/16	5/16	3/8
HS4.0	5/16	A	5/16	3/8	3/8	3/8
		B	5/16	3/8	3/8	3/8
		C	5/16	5/16	3/8	3/8

(Courtesy Square D Company)

TABLE 6.4. SUCTION LINE SIZES

Model	STD Suction Line	Arrange-ment	Distance Between Indoor and Outdoor Section			
			Up to 25′	25′ to 50′	50′ to 75′	75′ to 100′
HS1.5	5/8	A	5/8	5/8	5/8	5/8
		B	5/8	5/8	5/8	5/8
		C	5/8	5/8	5/8	5/8
HS2.0	5/8	A	5/8	5/8	5/8	3/4
		B	5/8	5/8	5/8	3/4
		C	5/8	5/8	5/8	3/4
HS2.5	3/4	A	3/4	3/4	3/4	3/4
		B	3/4	3/4	3/4	3/4
		C	3/4	3/4	3/4	3/4
HS3.0	3/4	A	3/4	3/4	7/8	7/8
		B	3/4	3/4	7/8	7/8
		C	3/4	3/4	7/8	7/8
HS3.5	3/4	A	3/4	3/4	7/8	7/8
		B	3/4	3/4	7/8	7/8
		C	3/4	3/4	7/8	7/8
HS4.0	3/4	A	3/4	7/8	7/8	7/8
		B	3/4	7/8	7/8	7/8
		C	3/4	7/8	7/8	7/8

(Courtesy Square D Company)

3. Use Table 6–4 to determine the correct suction line size. These sizes have been selected to provide the minimum pressure drop and to ensure proper oil return under most operating conditions.

For proper oil return in the suction line, refrigerant velocities must be maintained high enough to carry the oil along with the refrigerant back to the compressor. Minimum velocities should be 750 ft per minute for upflow runs. Line lengths should be limited to no more than 100 ft.

If the system is designed to maintain these velocities, there will be no need for traps in the suction line. Present practices in the building industry seldom allow sufficient space for traps in chases for tubing. It is more practical to size the line for proper velocity than to apply line traps.

Heat pump operation will have reduced refrigerant flow at lower ambient conditions; this factor has been considered in these sizing recommendations. The application of the accumulator ensures that oil will be returned to the compressor through the oil bleed hole in the accumulator.

7

COAL FURNACES

Coal-fired furnaces have been making a comeback in recent years (due to the rising cost of other fuels), and homeowners can save a great deal of time and money by doing their own troubleshooting and repairs when a problem occurs.

Troubleshooting coal-fired heating plants covers a wide range of problems, from checking for leaks in smoke pipes and flues to adjusting the automatic or manual controls. However, in nearly all cases, the trouble can be determined by using a systematic approach, checking one part of the system at a time.

Preventative maintenance is the best cure for all heating problems and if the hints found in this chapter are observed, operational problems will be kept to the minimum.

CHIMNEY KNOW-HOW

Since many problems with coal-fired furnaces are directly related to the flue and chimney, a review of good chimney maintenance is in order. Chimney flues should be of ample size and carried as nearly straight as possible from a point near the furnace to above the highest projection of the roof. Each flue preferably should be independent—having no connection with other flues or openings—and always be of the same area from top to bottom. A well-jointed tile flue with the joints tightly connected is compulsory, not only for proper operation, but for safety as well.

In looking over an existing chimney and in making connections to it, the following observations should be made:

1. See that there are no other openings into the chimney flue, either above or below the connecting smoke pipe.
2. If the division walls of the chimney contain more than one flue, it is best if they are carried up to the top of the chimney and down to the bottom of the chimney so that each flue is independent of the other throughout its entire length.
3. The area of the chimney flue should be maintained full size throughout its entire length and is free of all obstructions, such as loose brick, mortar, and the like, that might have become lodged in it. An offset in the flue, such as to go around an upper fireplace, should have increased area to overcome the added friction of the offset.
4. Look for an oily surface on the inside of the flue. This is an indication of creosote, usually caused by burning green wood. This substance can ignite and is often the cause of chimney fires. A mirror may sometimes be used to detect obstructions and the general condition of the flue.
5. The chimney should extend above the highest point of the roof or other immediate surroundings, such as adjoining buildings, hills, trees, and so on. If not, you can expect a poor draft.
6. The smoke pipe should not project too far into the chimney, as this will lessen the area of the flue at this important point.

To test for other openings or leaks in a chimney, twist a newspaper and light the end of it. Then hold it against the opening where the smoke pipe will enter. If part of the flame goes downward, as shown by C in Fig. 7–1, there is a leakage below the smoke pipe entrance. If the flame reacts as indicated by A in the drawing, there is a leak above the opening. A flame appearing as in B indicates that the flue is reasonably free of leaks, both above and below the opening.

In most cases, where the test indicates a leakage below the opening, it will be caused by two flues being joined together at the bottom for cleanout purposes, using only one cleanout door for two flues. New cleanout doors should be provided at the bottom of each flue and the existing (middle) cleanout door used to seal off the two flues with cement.

Where it is necessary to have the flue extend below where the stove's smoke pipe enters, unsatisfactory results may be avoided by giving the end of the smoke pipe the form indicated in Fig. 7–2. Care, however, must be taken to see that the pipe does not become turned the wrong way.

There are several ways to clean the flue of creosote and soot. Many people dangle chains down the chimney to loosen the debris and then use a special weighted broom to clean the chimney. If you do not have all the necessary

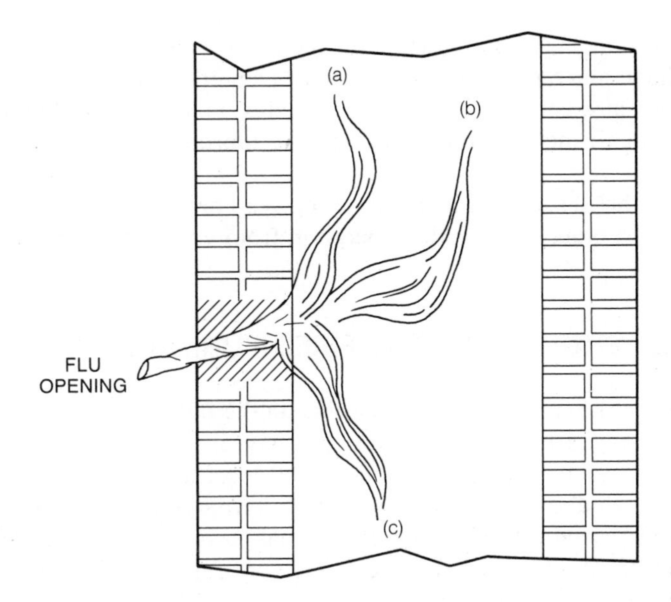

Figure 7–1. The condition of a chimney can often be determined by lighting a piece of twisted newspaper and holding it at the opening where the flue enters. The action of the flame will indicate conditions as described in the text.

tools, you will be money ahead by hiring a reputable chimney sweep in your area. Look in the classified section of the phone book under ''Chimney Cleaning'' or a similar entry.

If the flue joints are in need of repair, you can fill a canvas bag with sand so that is is tight enough to fit the flue, yet loose enough to slide up and down. Lower the bag by a rope or chain down to a point just below the joint that needs repair. Pour a small amount of mortar down the chimney which will rest on top of the canvas bag. Then work the bag up and down to squeeze it into the joint.

For best results, you will have to build up your chimney if it is less than 2 ft higher than the ridge of the house. You can follow the original pattern of the chimney or merely add a sheet metal extension.

Should your chimney have two flues, one should be higher than the other to prevent the draft from one flowing down the other. This should easily be

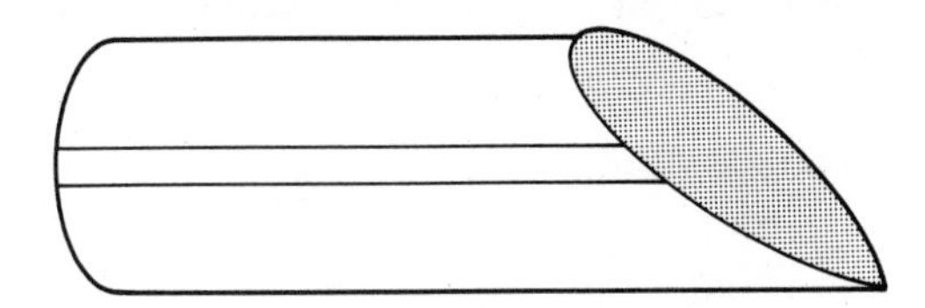

Figure 7–2. Method of cutting stove pipe for use where the flue extends below the entrance.

corrected by adding about 6 inches of flue tile to one of the flues, or you can install a hood and separate the two flues with a partition (often called a *wythe*). A hood will also correct downdrafts from hills, taller adjacent buildings, and the like, not to mention the protection the mortar joints will get from the elements.

With the items listed above in check, you are sure to have safer, more efficient, and more comfortable operation from your stoves or furnaces.

TROUBLESHOOTING COAL-FIRED FURNACES

Before firing the furnace for the first time in the fall, be sure that the furnace is clean. Also check the water level if you have a steam or hot-water system. See that the stoker is properly serviced and lubricated, and be sure that the feed screw, windbox, and tuyere openings are free of ashes, clinker, or other obstructions. Then fill the hopper with equal parts of fine and coarse coal to assure uniform feeding into the retort, and your furnace should be reasonably free of problems.

Assuming that your furnace has been properly installed and that it is of sufficient capacity, the following troubleshooting outline should help in solving many of the common problems. In this list the problem is noted first, then the possible causes of this problem are listed in the order in which they should be checked. Always do one and then observe the results before proceeding to the next. Finally, solutions to the various problems are given.

Problem 1: *There is insufficient heat.*

Remedy

1. The coal feed rate is too low; increase the coal feed.
2. The air supply rate may be too low; increase if necessary.
3. A poor draft can cause a problem; adjust the smoke pipe damper.
4. Check for a dirty fuel bed; clean if necessary.
5. The heating surfaces may be dirty; clean castings thoroughly.
6. Coal may be too fine; mix equal parts of coarse and fine coal.
7. The limit switch may be set too low; check the manufacturer's instructions to correct adjustment.

Problem 2: *There is too much heat.*

Remedy

1. Check controls for an incorrect setting; adjust to a comfortable setting.
2. The coal feed rate is too high; lower the rate of feed.

Problem 3: *There is excessive smoke in the chimney (during either the ON or the OFF period, or both).*

Remedy

1. Check the fuel bed, as it is probably too thick. If so, reduce the coal feed.
2. Not enough air may be causing the problem, so try increasing the amount of air to firebox.
3. Check for a poor draft and adjust the smoke pipe damper if necessary.
4. Coarse coal can cause excessive smoke in the chimney. Mix equal parts of fine and coarse coal.
5. Check for a faulty chimney or smoke pipe and repair if necessary.

Problem 4: *There is smoke in the furnace room.*

Remedy

1. The firebox draft is probably too weak; correct by adjusting the smoke pipe damper.
2. Check for too much air and if needed, decrease air by fan adjustment.

Problem 5: *Soot is accumulating.*

Remedy

1. The fuel bed is probably too thick; decrease coal feed.
2. The air supply rate may be too low; increase the amount of supply air.
3. The controls may be set incorrectly; increase operating periods of "Hold-Fire" control.

Problem 6: *There is insufficient air.*

Remedy

1. The fan adjustment may be set too low; increase the fan opening.
2. Check for siftings in the windbox; clean windbox if necessary.
3. Check for water in the air duct; remove the water.
4. The retort openings may be blocked; remove any foreign material.
5. The furnace room may be too airtight; allow more air to be admitted into the area.

Problem 7: *Gas is leaking into the hopper.*

Remedy

1. The fuel bed may be excessively dirty; clean the bed thoroughly.
2. Check for clinker in the retort; remove the clinker.
3. The fuel bed may be too thick; reduce the thickness of the bed.
4. The hopper lid may not be sealed; refit the lid and gasket.
5. The hopper smoke pipe may be plugged up; remove the cap and clean the pipe.

Problem 8: *Coke trees are incorrectly sized.*

Remedy

1. Heavy trees indicate that the feed is too fast; increase air or reduce coal feed.
2. Thin trees indicate too much air; increase the feed, decrease the air, or maintain a thicker fuel bed.
3. Coal may be too fine; mix coarse and fine coal.

Problem 9: *Deposits of fly ash can be seen.*

Remedy

1. The fuel bed may be too thin; increase the rate of coal feed.
2. Too much air can cause the problem; decrease the air.
3. Check for a dirty fuel bed; clean the fuel bed.

Problem 10: *The fire goes out.*

Remedy

1. Check the control circuit for operation and then rebuild the fire.
2. The clutch may be out or the pin sheared; throw in the clutch, replace the pin, or remove the obstruction.
3. Coal may be arched in the hopper; break the arch and push the coal down to warm.
4. The fuel/air ratio may be wrong; adjust one or the other.
5. Check for excessive draft; reduce the firebox draft.
6. The drive belt may be slipping; tighten, or replace a worn belt.

Adequate, regular lubrication ensures efficient operation of all movable parts, long equipment life, and minimum maintenance cost—but never overlubricate. Observe the manufacturer's instructions when lubricating the various parts.

8

WOOD AND WOOD FURNACES

Since the supposed liquid-fuel shortage of the mid-1970s, many homeowners have gone to wood fuel, either as their sole source of heat or as a supplemental heat. This is fine provided that one has a source of either free or relatively inexpensive firewood; otherwise, the wood will probably cost more than conventional fuel oil and natural gas. Many suburban residents with a woodlot or other means of obtaining and curing good firewood can earn a good five-figure income by hauling the wood to cities and selling it for fireplace wood. It is nothing for an urban resident to pay as much as $80 for a pickup truck load of wood, and this is not usually a full cord. If a wood furnace is used to heat an average home of, say, 1500 ft^2, it would take at least a cord of wood per week, maybe more.

Of course, in recent years, smaller wood-burning stoves have appeared on the market by the thousands. All are capable of heating a single room or area, and if the right airflow system is arranged, such a stove can keep the living area of the average home relatively comfortable; many families are using them for 90% of their heating needs. However, there are also semiautomatic heating plants (wood furnaces) available that usually start at $1500 and up for an installation. Although there are different types, most operate very similarly to the conventional coal-fired, forced-air furnace; that is, wood is placed in the firebox of the furnace, ignited with kindling or an automatic spark igniter, and allowed to burn. The box must be loaded with enough fuel to last the desired or prescribed time (the average being from 12 to 16 hours). A damper, controlled by a thermostat in the living area of the home allows the furnace to get hotter or cooler, and when the thermostat calls for heat, the fan/coil starts to force the heated air through ducts to the living area.

The heat-transfer medium can be water, just like a conventional hot-water heating system described in Chapter 2. Instead of the fan/coil unit and plenum, a coil circulates water in and around the walls of the firebox, where it is heated. When the room thermostat calls for heat, a solenoid valve opens and energizes a pump to circulate the heated water through the piping system and radiators in the home.

In addition to the difficulty of finding reasonably priced wood in some areas, this type of furnace is also not the cleanest type of heat available. There is frequent maintenance required, removing ashes, and the entire home will also have to be cleaned more frequently if the furnace is located in the basement. However, the latter problem has been somewhat overcome by locating the wood furnace some distance from the house in a fireproof, masonry outbuilding. The wood supply can also be located in this spot, and with such an arrangement, there is less operation and maintenance—making the firing of this type of furnace almost a pleasure—especially since you realize the amount of money you are saving.

Here is some of the necessary know-how for operating a wood-burning furnace and for getting the most out of such a furnace.

Successful burning of volatiles requires oxygen, a long flame path, and a temperature of about 1000°F. Most of the older wood-burning units do not make provisions for this, but allow gases to escape up the chimney, where their heating value is lost. Better designs are needed before higher efficiency can be obtained.

If all the primary air for combustion is admitted at the base and under the burning wood, most of the oxygen reacts in the bed of glowing charcoal to produce carbon monoxide, carbon dioxide, and heat. The heat rising through wood above the charcoal bed promotes the release of volatiles, but these cannot burn since most oxygen in the primary air has been depleted in passing through the bed of coals. So the unburned volatiles pass up the chimney, sometimes dissolving in condensed steam on chimney surfaces to form pyroligenous acid and eventually, a gummy deposit called *creosote.* Acids cause corrosion of iron and steel, while deposits of creosote may ignite and fuel dangerous chimney fires.

Improvements of efficiency can be obtained by designing secondary air inlets so that oxygen can be introduced and mixed with unburned volatiles without cooling them below their ignition point; complete combustion can then occur. Industrial wood-fired boilers are designed to blow secondary air in domestic wood stoves.

GOOD WOODS

Table 8–1 lists the fuel value of the most common trees found in the United States. Each group is arranged in order of descending fuel value. In the left-hand column, you can see that live oak is better than black locust. But black locust is better than dogwood, and so forth. For more information, see Table 8–2.

There are certain characteristics of which users of wood-burning appliances should be aware. For example, the bark of white birch gives off much soot while burning. Oaks have excellent fuel value, as does black walnut, but give off a pronounced acrid odor. Other types are apt to harbor insects and should be used up before warmer weather.

All evergreens, except red cedar and Douglas fir, should be used only in a pinch. The softer hardwoods, such as basewood, yellow popular, butternut, bit tooth aspen, and cottonwood, are not as satisfactory as the denser kinds, such as maple, oak, hickory, beech, yellow birch, and various fruitwoods.

GATHERING FIREWOOD

If you have a woodlot consisting of good hardwood trees, you are more fortunate than most, as firewood is becoming a valuable commodity. If not, there are still other sources for free firewood; that is, free for the cutting. For ex-

TABLE 8.1. FUEL VALUE OF COMMON AMERICAN TREES

Best	Okay	Only in a Pinch
Live oak	Holly	Hemlock
Shagbark hickory	Pond pine	Catalpa
Black locust	Loblolly pine	Yellow poplar
Dogwood	Shortleaf pine	White spruce
Hophornbeam	Red maple	Black willow
Persimmon	Cherry	Bit tooth aspen
Apple	American elm	Butternut
White oak	Black gum	White pine
Honey locust	Sycamore	Balsam fir
Black birch	Douglas fir	Cottonwood
Red oak	Pitch pine	Basswood
Sugar maple	Sassafras	
American beech	Magnolia	
Yellow birch	Red cedar	
Longleaf pine	Bald cypress	
White ash	Chestnut	
Black walnut		

TABLE 8.2. GENERAL DATA ON COMMON AMERICAN WOODS

Species, listed in approximate order of density	Available heat per cord (million Btu)		Percent More Heat From Air-Dry	Price/cord[a] Equivalent to Heating Oil at \$0.50/gal[b]		Price/cord Equivalent to Electric Baseboard Heat at \$0.04/kWh[b]	
	Green	Air-dry		Average fireplace	Airtight stove	Average fireplace	Airtight stove
Hickory	20.7	24.6	19	\$20	\$68	\$43	\$144
White oak	19.2	22.7	18	19	62	40	133
Sugar maple	18.4	21.3	16	18	59	37	125
American beech	17.3	21.8	26	17	60	38	128
Red oak	17.9	21.3	19	18	59	37	125
Yellow birch	17.3	21.3	23	18	59	37	125
Yellow pine	14.2	20.5	44	17	56	36	120
White ash	16.5	20.0	21	16	55	35	117
American elm	14.3	17.2	20	14	47	30	101
Red maple	15.0	18.6	24	15	51	33	109
Douglas fir	13.0	118.0	38	15	49	32	105
Eastern white pine	12.1	13.3	10	11	37	23	78
Aspen	10.3	12.5	21	10	34	22	73

[a]A full cord is a stack of wood 4 ft high by 4 ft wide by 8 ft deep.

[b]Assumed efficiencies: fireplace, 15%; airtight stove, 50%; oil furnace, 65%; electric baseboard, 100%.

ample, the U.S. Forest Service opens certain areas for cutting firewood for one's own use. You cannot sell it, but you are allowed to cut a certain amount for your own use as fuel. Some areas have been closed to the public, however, because of abuse of this privilege.

Another source is right-of-way clearance for power lines or highways. When such construction is taking place, the power company or the highway department will usually be happy for you to cut all the wood you want in the areas to be cleared. You will be helping them clear rights-of-ways in exchange for free firewood.

You may also try some of the local farmers in your area. Many will be glad to sell you firewood cheaply, especially if you will do the cutting and hauling.

If you intend to cut your own firewood, early spring is the best time. Weed growth is not too great then and the wood will have all summer and early fall to season.

If you live near a river or stream, driftwood is always available, especially in early spring after the high-water level has subsided. Driftwood comes in all types, sizes, and conditions. Small, dry sticks are fine for starting a fire, while larger, moister ones are good for banking a fire. It can usually be found on the outside bank where the river or creek bends sharply.

Driftwood used for fuel does have its drawbacks. The most difficult part may be hauling it out. Unless there is a suitable road directly to the location, you will have to carry it out on foot or use a boat to haul it to an access point. Another disadvantage of driftwood is that it is likely to be contaminated with sand and silt, which will dull your saw or ax blade very quickly. Those pieces gathered from salt water can also carry corrosive chemicals into the stove. Depending on the type and thickness of metal used in your stove, this corrosive material can cause portions of your stove to rust out in a much quicker time than it normally would.

PAP-O-FIRE

So you don't have any wood at all? You can build a pap-o-fire in your fireplace and still have lower heating bills, or so says Dwight Mills of Fredericksburg, Virginia.

Mills—now retired—has worked in the heating field most of his life and has devised a neat, simple method for transforming old newspapers and used crankcase oil into "free" heat for the home. You take old newspapers and roll them up until they are about 4 in. in diameter. They are then held together with a single piece of reinforced packaging tape—the tape that comes on rolls and is practically impossible to cut. These paper logs are then stood up in a tin can in which used motor oil has been poured so that each roll of paper soaks up the oil.

These oil-soaked paper logs will give out good heat just like they are, but for more efficient burning, Mills has designed a special two-level grate to provide sufficient air for complete clean combustion of the oiled newspapers. For greater heat radiation, Mills uses three or four steel logs (or cylinders) on the grate's upper level, in addition to three oil-soaked paper logs. He also uses one real wooden log to complete the combination. Then, as the fire burns, the steel logs become hotter and hotter, which provides the air entering the fire to be preheated before it begins combustion. The result is a hot, clean, nearly pollution-free blaze that leaves practically no ash residue.

The total cost of the pap-o-fire setup is less than $100. For more information, you can write to Dwight Mills, 1500 Winchester Street, Fredericksburg, Virginia 22401.

You can also find newspaper rollers in inexpensive gift catalogs for only a few dollars. They will enable you to consistently make tightly packed paper logs for use as fuel in your fireplace or as an aid in starting fires.

HARVESTING WOOD

Let's assume that you have found your supply of firewood; now you have the problem of getting the wood from tree to stove in a safe manner. Your best bet, and often the desire of the land owner, is to look for downed trees; those that have been downed by high winds, ice storms, and the like. This makes the chore easier and helps to clean up the woodlot.

The first step in cutting up a downed tree is to cut off the limbs approximately flush with the trunk. This operation is called *limbing.* A sharp axe is about the only tool that you will need except on the larger trees, where a chain saw or bow saw will make the work easier and faster. Start on the free limbs (those not supporting the weight of the tree) working from the base of the trunk toward the top. The first cut on each limb should be made on the side facing the base of the trunk, that is, the underside of the limb when the tree was upright. Cut and trim all the free limbs as close as possible to the trunk.

Now, the tricky part: the cutting of the supporting limbs—the ones that are supporting the trunk. A wrong cut in the wrong place can cause the tree to come crashing down to the ground, pinning your ax or chain saw with it. The object is to lighten the trunk before cutting too many of the supporting limbs. Depending on the type of tree and several other factors, you should usually start at the top end of the trunk when cutting the supporting limbs. The tree is first topped and then the next couple of supporting limbs are chosen and cut, starting the cut on the outside of the bend a few inches away from the trunk. When the limb is free, the stub is cut flush with the trunk. Cut buck or cut the trunk into sections after measuring for the correct length for your stove. The trunk is cut in sections back to the next supporting limbs. Again, a few more supporting limbs are chosen and cut away from the trunk. One must be careful to judge the role of each supporting limb as far as its support is concerned. The wrong cut and the entire trunk can turn and roll on you or knock your chain saw into your leg. When making such cuts, make certain that you have adequate room to retreat should the tree start to roll or fall. Be on guard at all times during the cut, and be prepared to pull the chain saw away from the cut if you feel the weight of the tree pinching the guide bar. It is wise to have an ax or second saw ready in case the saw binds during the cutting; a second cut farther from the trunk will usually release a bound tool. Leave a few of the supporting branches to support the trunk during bucking.

Cutting the trunk and major branches into stove-length sections can be done with either a chain saw or a bow saw. Using an ax can be time consuming and hard work. With a little experience, most people can "eyeball" the sections and cut to the right size without measuring, but at first it is best to measure each section, then mark by scoring with a light saw cut or ax. Working from the tip of the tree toward the base, cut the trunk into sections as marked, and the remaining supporting branches as you get to them. The major branches can usually be laid across the trunk or another section of wood and sawed on the spot. However, a sawbuck will make the job easier.

In bucking the trunk, the position of the piece being cut will dictate your first cut. If the length is free (not supported by a branch) and not extremely large, start on the top side of the section and cut right through with the chain saw. If the section is large, make an undercut about one-fourth the thickness of the section and then finish the cut from the topside. When doing so, however, be careful not to let the section drop on a toe or foot.

If the section is supported by a branch, the cutting procedure is reversed; that is, make the first cut from the top halfway through the thickness of the section, then finish the cut from the bottom. If the cut is made completely through from the top side, the supported piece can buckle inward and pinch or bind the saw.

When the supporting limbs have been removed, you may still have a section of trunk to cut up. Since no supporting limbs are present, the trunk will be resting directly on the ground. Trying to cut through the trunk in this position, you will surely cut into a rock and dull your saw blade. Make all cuts about halfway through the thickness of the trunk. Then roll the trunk over and finish these cuts from the opposite side.

The sections of wood are now ready to be hauled to your home or place of utilization. Try to pick a place to store the wood that will not be unsightly, yet close enough so that you won't have to walk far to carry the wood.

SPLITTING

Splitting firewood can be fun if you have the proper tools and go about the task in the right way. It can also be a back-breaking experience if you don't have the right tools or don't know what you're doing.

Green wood splits much more easily than seasoned wood, so it is a good idea to split the wood as soon as possible after cutting it. First separate the various pieces into size groups. Usually, any piece under 6 in. in diameter need not be split. Six- to 8-in.-diameter pieces should be set aside to be split with an ax. Pieces larger than 8 in. will normally have to be split with a splitting maul. Extremely large or difficult-to-split pieces may require a sledge hammer and two or more wedges. A power splitter will also do the job, but are expensive, and unless your health dictates otherwise, it is good exercise.

The technique of splitting takes a little getting used to, but once you get the hang of it, it is not difficult at all. First, you want to provide some support for the piece being split and prevent your ax from striking the ground. A medium-sized log will do. Also, make sure that you have the proper-size handle on your ax and know the feel of it during the swing. A handle that is too short for you can let the blade come back against your foot or leg.

The piece to be split should rest against the support log or chopping block in a vertical position. If the log has been lying around for a while, the ends of the log will have started drying out. However, when the log is bucked, you will have one "new" or greener end on each piece to be split. This "new" end should be facing upright for splitting; the section will split easier in this position. In soft ground, you may also want to place a board under the other end to keep it off the ground. If the ground is too soft, it will absorb much of the energy of the ax swing, and if the ax should cut into the dirt, it will become dull very quickly.

If the piece is small, aim the ax blade at about the center of the section and let go, giving the ax handle a slight twist to rotate the blade the moment the blade comes into contact with the wood. Look for radial cracks on the larger pieces and aim for them.

The novice can expect to get the ax stuck especially if he tries to take off too big of a hunk. The ax will bind in the split, as your blow will not entirely cut through the piece of wood. When this happens, pound a wedge in the crack to release the ax. You can avoid binding your ax by not cutting large pieces of wood directly in half. First split away the edges until you have reduced the piece to a reasonable size; then split through the remaining hunk.

If your chain saw is on the blink, you can cut some firewood for immediate use with the ax alone. You can fell the trees by chopping and then limb the tree as described previously. Instead of chopping each piece of firewood in the desired length, use a maul and several wedges and split the entire trunk lengthwise. If the tree is relatively small, one split will usually be sufficient. On larger trees, you will first want to split the trunk in half, and then into quarters or possibly eighths. With the cross-sectional area reduced in size, the individual pieces will then be easier to chop into usable lengths. This is not the recommended practice since bucking a trunk with a saw is much easier, but in certain cases, this could mean the difference between staying warm or being cold.

STACKING AND SEASONING

With your supply of wood cut and split into the proper-sized pieces, it must be stacked in a way to allow the wood to season properly. The time required for seasoning wood varies according to the species, but from 6 to 9 months is usually sufficient to dry most woods. Soft woods such as pine should not be stored for much longer than a year or 18 months because they will begin to decay after this time and lose much of their heating value. Hardwoods, such as oak, hickory, and the like, will normally last up to 2 years of storage before they start to decay. You shouldn't have more than a 2-year supply on hand at any one time.

For wood to season to the ideal dryness, it should be off the ground and sheltered from the weather. Two 12-ft 2 × 4s laid in parallel on the ground will act as a support, as will parallel poles or a stringer of concrete blocks. When using wood, the poles should be treated with a wood preservative to prevent rot. With the wood elevated above the ground, air can circulate freely under the pile and speed drying.

Wood can be stacked in two different ways: parallel for saving space and criss-cross for drying faster. Most stacks of wood should not be higher than 4 or 5 ft, and if much more than 2 or 3 ft, you will have to brace the ends.

Position the wood so that the bark side is up to repel water—even if the wood is stacked under cover—since some rain will probably blow into a lean-to or similar shelter and spray the wood. If the wood is completely under cover, such as in a garage, it does not matter how it is stacked.

WOODSHEDS

One of the simplest "woodsheds" is to drive four stakes in the ground (two at each end) and provide parallel poles between them for the wood to lie on. To protect the wood from the weather, you can spread a sheet of plastic material over the pile and anchor it in place with rocks or pieces of wood.

A little more sophisticated storage idea is to make end frames. Since these frames are movable, you may make your woodpile as large or small as you like by merely moving the ends inward or outward to accommodate the amount of wood. The weight of the logs on the bottom 2 × 4s of the end frames anchor the frames at the ends of the stack. When the wood diminishes, you just move the end frame in closer and stack more wood on it to anchor it and to get the remaining wood tightly stacked.

9

AIR DISTRIBUTION SYSTEMS

Air distribution systems carry the air from the blower unit to the space to be conditioned, and then back to the furnace or blower for redistribution (Fig. 9–1). In designing and installing such a system, many calculations must be made as well as decisions to enable the proper size of equipment to be used, and also to select duct routes, outlets, and controls that will result in an adequate system.

Air ducts are available in a variety of shapes and sizes, but the ones most used in residential work will be either circular, rectangular, square, or a combination. The size of each duct system will depend on the amount of air to be carried through them.

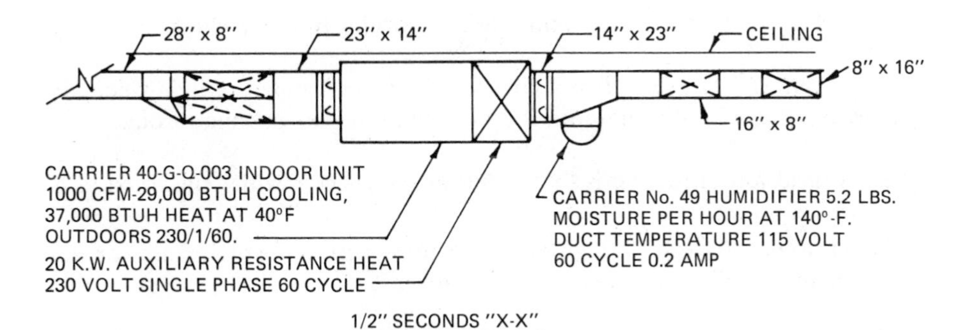

Figure 9–1. Typical air-handling unit used in a residential heating/cooling system. (*Courtesy of the author*)

Duct systems are divided into two general categories: those designed for areas in which the cold season is the major consideration, and the other for areas in which the hot season is the major consideration. Generally, duct systems that supply air at or near floor level are best for areas where the heating system is the major concern; this, of course, is due to the fact that warm air rises. Overhead systems are normally used where the cooling system is the major concern since cool air falls. Of course, there are exceptions to the preceding statements. It is often not practical to install ductwork from below because of the building construction, economy reasons, and the like. The reverse can also be true; that is, it may be impractical to install the ductwork overhead because of the ceiling structure or for a similar reason.

SUPPLY-AIR OUTLETS

Supply-air outlets are a major part of any air distribution system because they provide a means of distributing properly controlled air to a room or area. To accomplish this goal, the supply-air outlets must deflect or diffuse the air, be adjustable for changing the airflow rate, void of air noise, and be able to throw the conditioned air no less than three-fourths of the distance from the outlet to the opposite wall.

Many types of supply-air outlets are available, a sampling of which follows:

Supply-Air Outlet: A supply-air outlet can be a floor, ceiling, or wall opening through which conditioned air is delivered to a room or area.

Ceiling Diffuser: The ceiling diffuser is a square, oval, circular, or semicircular facing device that covers the supply-air opening of a room or area. Most diffusers are adjustable for airflow direction and rate.

Grille: The grille is a covering for any opening through which air passes.

Register: The register is a covering for any opening through which air passes and has a built-in damper for controlling the air passing through it.

1. A fixed-louver register is a nonadjustable register in which the air pattern is factory set.
2. An adjustable-louver register is a register with adjustable bars for directing the air in several different patterns.

Figure 9–2 shows several types of supply-air outlets and how they would appear on a working drawing.

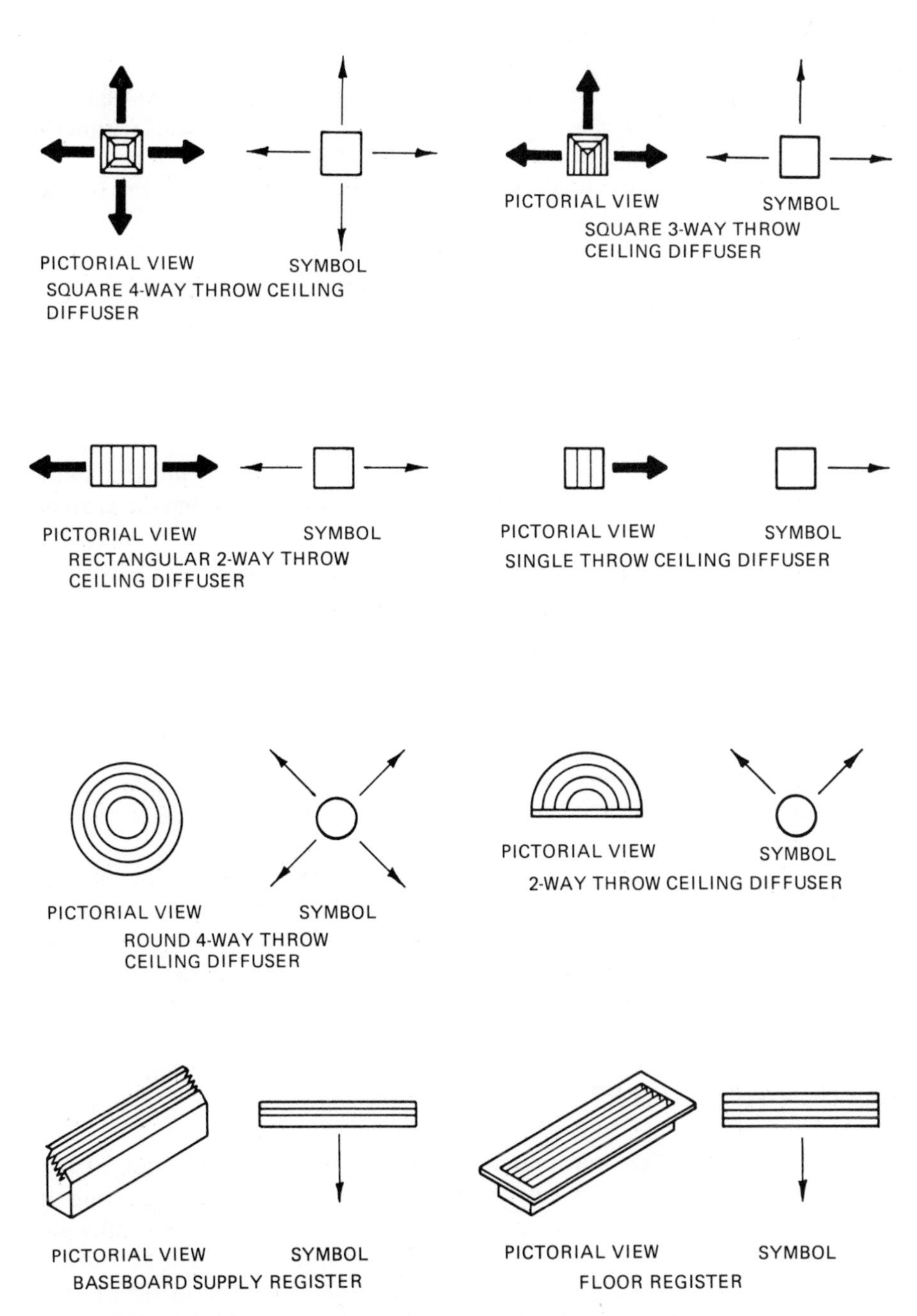

Figure 9–2. Several types of supply-air outlets and how they appear on working drawings. (*Courtesy of the author*)

LOCATION OF SUPPLY-AIR OUTLETS

Supply-air outlets in the area to be conditioned must be located so that sufficient air is supplied to establish and maintain comfort conditions within the area: that is, a uniform air pattern that is free of hot or cold drafts.

The location of supply-air outlets is determined after taking the following into consideration:

1. The size and shape of the room or area to be conditioned
2. The decor of the area to be conditioned
3. The furniture placement within the area to be conditioned
4. The ceiling, wall, and floor finish of the area to be conditioned
5. The budget required
6. Whether the system is designed for heating, cooling, or both (if both, the one that is used the most during the year should be given the greatest consideration)
7. The total air quantity required for the area under consideration
8. The total load and draft conditions of the area
9. The diffusion of spread pattern required within the space to be conditioned
10. The level of noise allowed in the space to be conditioned

Therefore, supply-air outlets are located according to the shape, size, usage, and load concentration of the area, and according to whether the system is used in a cooling or a heating application or both.

In all cases, drawings of the areas should be made. These drawings should be studied to determine the size, shape, and use of each room or area. If architectural drawings of the home are readily available, these are excellent. However, if none are available, scale drawings will have to be made by measuring the various areas and then transferring these measurements to paper.

RETURN-AIR OUTLETS

The location of the return-air outlets is not as critical as the location of the supply-air outlets, but as a rule of thumb, the return-air outlets should be in the opposite location from the supply-air outlets. That is, if the supply-air outlets are located in the ceiling or high on a side wall, the return-air outlets should be located in the floor or low on a side wall. If the supply outlets are located on the outside wall of the room, the return-air outlets should be located on an inside wall of the room, preferably on the opposite wall. If, for example, the supply outlets were located in the ceiling and the return outlets were also located in the ceiling, there is danger of short-circuiting the air from the supply outlets directly to the return outlets, as shown in Fig. 9–3.

If all supply-air outlets are located around the perimeter of a building and the air within the building is not "sealed" in each space, one centrally located return outlet is usually satisfactory. Transfer grilles, undercut doors, or a ceiling plenum can be used to ensure that all supply air returns to the one return outlet.

The return-air system for many multistory residences is located in the stairway, due to the fact that the cooler air flowing down the stairs will flow into the return outlet in the stairway; this minimizes cold-floor drafts.

When conditioned space must be "sealed off," preventing one central-return outlet, the return-air outlets must be located in several areas and then transferred by duct to the air-handling apparatus.

AIR-HANDLING UNITS

Fans are used in air-handling units for the circulation of air in a heating system. They are manufactured in the four general types, illustrated in Fig. 9–4. Each type of fan shown can have either a belt drive or a direct connection.

Centrifugal: The widely used centrifugal fan, in which the air flows radially through the impeller, can efficiently move large or small quantities of air over an extended range of pressures; for this reason, it is the most versatile type in use today.

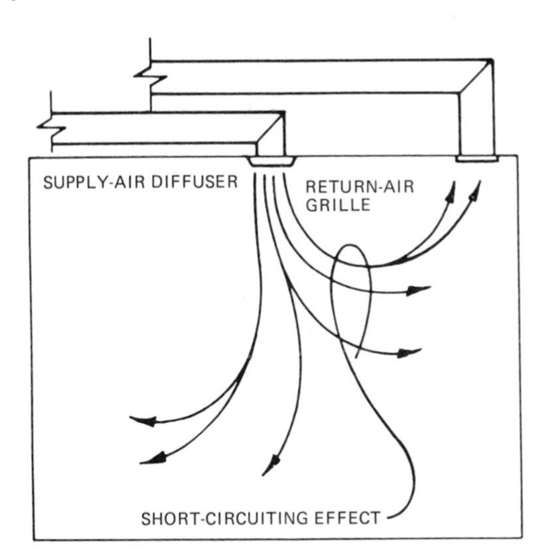

Figure 9–3. Example of short-circuited air caused by placing the supply and return outlets too close together. (*Courtesy of the author*)

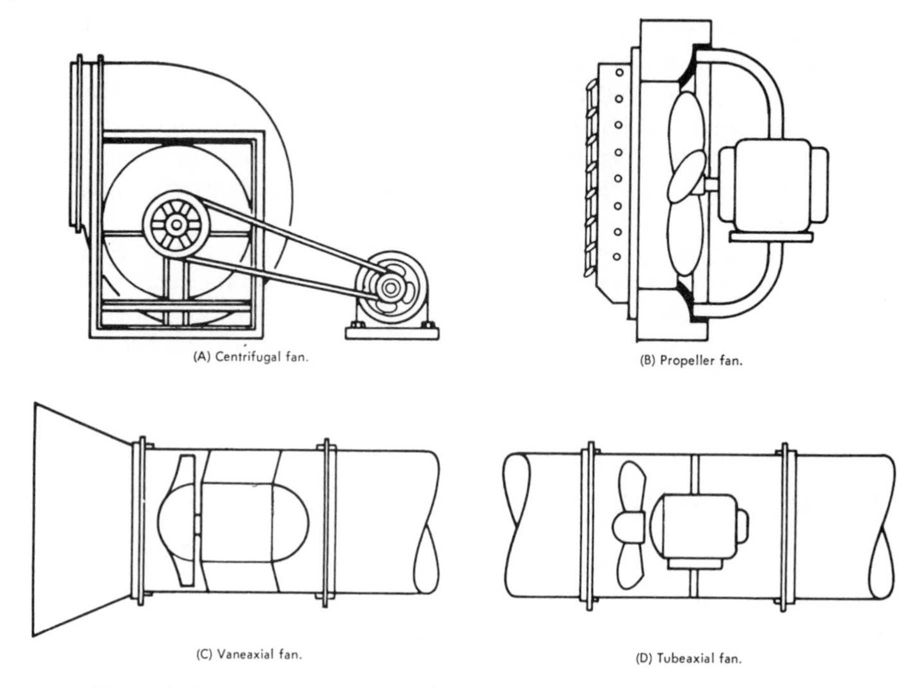

Figure 9–4. Four general types of fans used in residential heating system. (*Courtesy of the author*)

Propeller: The propeller fan can move large quantities of air, but its use is limited to areas where there is no duct system and where the resistance to airflow is low.

Vaneaxial: The vaneaxial fan produces an axial flow of air through the wheel and blades. It is capable of moving large or small quantities of air over a wide range of pressures. This type of fan is always used in a duct system since the wheel and its blading are located in a cylindrical housing. Air-guide vanes are used either before or after the wheel.

Tubeaxial: The tubeaxial fan is similar to the vaneaxial fan in that the air flows axially through the impeller. It, too, is capable of moving small or large quantities of air over a wide range of pressures. However, most axial fans produce more noise than centrifugal fans do and are therefore limited to use in areas where noise levels are of secondary concern.

FAN-COIL UNITS

Air-handling or blower units are used in conjunction with a heating source to move the air from the heat source to the rooms or areas to be heated. For

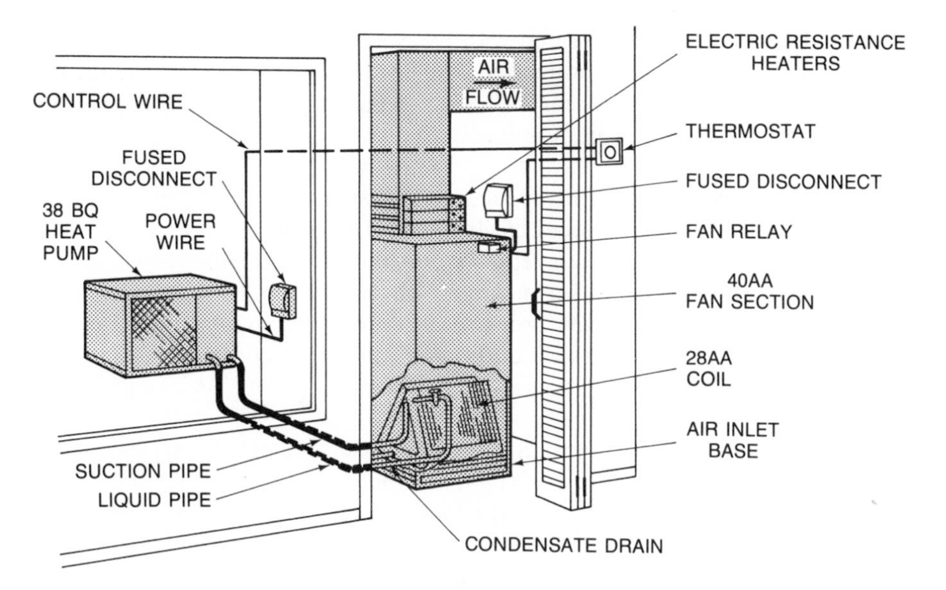

Figure 9–5. Plenum and ductwork arrangement for an up-flow residential heating/cooling system. (*Courtesy of the author*)

example, in an oil-fired furnace, a blower unit moved the heated air through the ductwork.

Figure 9–5 shows an up-flow furnace with the cooling coils on the return-air side of the furnace, which is used in conjunction with the heater to offer cool air in hot months and hot air in cold months.

FIBERGLASS DUCTWORK

The majority of ductwork used in residential heating systems is of the sheet-metal type, but fiberglass ductwork has been used for a couple of decades with good results and is the preferred type for installation by homeowners.

Unlike ductwork constructed from sheet metal, fiberglass duct requires only a few common hand tools for assembly and installation: a sharp knife, a hammer, pliers, a heat-seal tool, butt-joint tools, and a staple gun.

Furthermore, a conventional sheet-metal system requires consideration of thermal efficiency, acoustical efficiency, and vapor resistance as well as air delivery. All these features must be given careful, time-consuming consideration. Fiberglass ductwork has insulation, sound absorber, and a vapor barrier built in.

Round Duct: Prefabricated round duct is quickly and easily installed with standard sheet-metal fittings and the other tools mentioned previously. The sheet-metal fittings assure proper alignment and provide reinforcement at the joints, which are also secured with pressure-sensitive aluminum tape and heat-sealed. This assures continuation of the vapor barrier and strong, airtight joints. Should a puncture occur in the vapor barrier, it can be mended easily by applying tape over the damaged area.

Rectangular Duct: Rectangular fiberglass duct is easy to use: even the largest sections can be positioned easily due to their light weight and resilience. Sheet-metal connectors are not required for joining rectangular duct sections unless the duct span exceeds the maximum allowable span for a specific thickness of duct board at a given pressure. However, rectangular duct runs do require suspension on 6-ft (maximum) centers unless the duct can be supported by structural members.

Combined Duct Installations: Round duct take-offs from rectangular duct are easily accomplished with fiberglass duct. A round sheet-metal fitting the same diameter as the round duct take-off is secured to the rectangular trunk. The round take-off is then slipped over the metal fitting and pushed tightly against the rectangular trunk, taped around the curvature of the connection, and then heat-sealed.

The exact size of duct to use depends on the amount of air to be delivered to each room. Although the design of duct systems is beyond the scope of this book, here is basically how it is done. Once the amount of heating has been calculated for each area of the home, a unit is selected that will deliver this amount of air. The specifications of the unit will also give the total cubic feet per minute (cfm) of air the furnace or air blower will deliver. Since the cfm of the heated air is directly proportional to the amount of heat delivered, the branch ducts are sized to deliver the given amount of air to each area at a velocity that will not cause unnecessary noise. Duct-sizing calculators are available that will simplify this procedure. One, distributed by Carrier, works on the order of a circular slide rule. A chart in the appendix of this book will suffice for most residential calculations.

10

INSULATION REQUIREMENTS

Few items in the home cost more to operate than heating and cooling systems, especially in the northern states where winter temperatures drop to zero or below. One way to beat the spiraling cost of fuel is through insulation of ceilings, walls, floors, ducts, pipes, the water heater, and other items in the home that stores heat. Caulking, weather stripping, and installing storm windows and doors finishes off the job.

Winter heat loss occurs through the ceiling, walls, and floors, but mostly through poorly insulated windows and doors, not to mention the infiltration through cracks around windows and doors and other places in the structure. A significant amount of infiltration occurs around electrical outlets where insulation is thin or nonexistent.

Of course, some air changes in the home are necessary to maintain a healthful environment. However, infiltration should be slowed down to a degree that will provide a healthy atmosphere rather than the excessive level of infiltration found in many homes today.

The amount of insulation needed in various areas of the home depends a great deal on the location of the home, and its value is expressed as an R-value, an R followed by a number. The R-value of an insulating material is the most reliable indication of the job it can do when the insulation is properly installed, and this value indicates the material's ability to resist the flow of heat passing through it. In general, the more resistance, the higher the R-value and the better job the material can do.

Some of the more commonly used insulating materials are as listed below. Each has a specific application. The number of inches needed for each will depend on the R-value desired or that which it is possible to install.

Batts: Batts currently are available in glass fiber and rock wool and are used mainly to insulate unfinished attic floors, unfinished attic rafters, the underside of floors, and open sidewalls.

Blankets: Blankets of insulation are similar to batts except that the material comes in rolls instead of batts. The material is again glass fiber and rock wool, and their use is identical to that of batts.

Formed-in-place: Expanded urethane can be blown into finished frame walls by first cutting a small hole (about 1 in. in diameter) and then blowing or pumping the material into the void areas through a hose and nozzle. When filled, the holes are plugs and finished to match the existing wall finish.

Rigid Board: Such materials as expanded urethane, glass fiber, and polystyrene are pressed into rigid boards and used to insulate exterior walls before the final siding or masonry is applied. These boards are also used quite extensively on basement walls and around the floor perimeter when a concrete slab is poured.

Loose Fill Blown In: Fiberglass, rock wool, and cellulose materials can be blown in and used in such areas as an unfinished attic floor, finished attic floor, finished frame walls, and the underside of floors when a backing board is used to hold the material in place.

Loose Fill Poured In: The same material as that described in the paragraph above can also be used to insulate unfinished attic floors. Vermiculite is also used to insulate the hollow areas of masonry block.

Attic spaces in both new and existing homes should be insulated with a minimum of 4 in. of insulation. Six inches is better and many homes have up to 10 in. especially those utilizing electric heat. If the attic space is unfinished and readily accessible, glass fiber batts or rolls can be placed between ceiling joists, or loose fibers can be poured into the spaces. However, many homes have had an attic floor installed, which prevents using batts or poured insulation. In cases of this sort, blown insulation is the most practical application (Fig. 10–1). Removing a few floorboards and using an insulation blower will enable the space to be filled between the attic floor and the ceiling below. Of course, floor boards may be removed and regular batts may then be installed prior to reinstalling the floor boards, but this can be quite time consuming, especially if there is much "junk" on the attic floor.

Most insulation dealers will loan or rent small portable insulation blowers when insulation is purchased. Most will also provide complete instruction

Figure 10–1. Installing cellulose fiber loose-fill insulation in attic. (*Courtesy of Pal-O-Pak*)

on the blower's use. When using a blower the operator should wear a respirator, as the fibers are harmful to the lungs.

New homes are normally insulated by insulation board on the outside of the studs (between the studs and the outside finish) together with 4-in.-thick batts of fiberglass insulation between the studs (on the inside of the fiberboard insulation). In existing homes with wood siding, one of the top siding boards is removed around the perimeter of the house, and loose insulation is either poured or blown into the space between the studs (Fig. 10–2). Alternatively, small holes can be drilled through the siding at the top of each between-stud cavity to accommodate the nozzle of the blowing machine and tubing. Then the siding is carefully replaced or wooden plugs are used to plug the holes before the surface is refinished.

If the house has masonry outside walls, insulating the walls becomes more of a problem, but it can be done; the exact method used depends on the structure of the home and other circumstances. For example, if the interior walls are drywall, the inside of the outside walls can be drilled and the insulation blown into the voids between the studs via these holes. Other installers may find it best to drill down through the partition plate to gain excess to the stud area. However, since most modern homes have bridging or fire stops installed about halfway down between studs, blown insulation installed in this manner will not completely fill the space; only the top half can be filled from above. To fill the lower area, however, the baseboard can be removed along the exterior wall, holes drilled in each partition, and the insulation blown in

Figure 10–2. Installing cellulose fiber loose-fill insulation in sidewalls of home by removing boards at the top and bottom. The installation can also be done by drilling 1-in. holes at the top and bottom of each void area and using the nozzle to blow in the insulation. (*Courtesy of Pal-O-Pak*)

this way, or perhaps the insulation can be blown up from the basement. Of course, all holes should be filled or patched once the insulation has been installed.

Fiberglass batts are best for use in unfinished attics and for unfinished underfloors. Purchase the type with backing and of the correct width to fit perfectly between the joists or rafters. Then use a staple gun to secure them in place. Table 10–1 shows various types of house construction and how they should be insulated.

VAPOR BARRIERS

Batts and blankets come with and without vapor retarders attached in the form of kraft paper, often sprayed with asphalt or layered with polyethylene or foil. Rigid boards are impermeable, but loose fill often needs a vapor retarder added.

An acceptable effective moisture-retarding material is any material with a perm rating less than 1. The rating is a measure of a material's permeability or characteristic of allowing moisture vapor to pass through.

INSULATION FACT SHEET

THIS IS FOR PAL-O-PAK CELLULOSE FIBER LOOSE FILL INSULATION

COVERAGE CHART

PLEASE COMPARE INSULATIONS BY USING THE NET COVERAGE FIGURES WHICH ARE HIGHLIGHTED IN YELLOW. The figures showing joists are provided only to give you a fair comparison with fact sheets that do not show net coverage figures.

R Value	Minimum Thickness	Maximum Coverage			Maximum weight per square foot
To obtain a thermal resistance (R) of:	**Installed Insulation should not be less than:**	**Type of area to be insulated (with or without framing joists):**	**Maximum sq. ft. coverage per 15 lb. bag:**	**Bags per 1,000 sq. ft.: (15 lb. bags)**	**The weight per sq. ft. of installed insulation should not be less than:**
BLOWN ATTIC					
R-40	10.9 " thick	NET COVERAGE (no joists)	7.18 sq. ft.	139 bags	2.09 lbs. per sq. ft.
		2"x4" joists/16" center	7.50 sq. ft.	133 bags	2.00 lbs. per sq. ft.
		2"x6" joists/16" center	7.66 sq. ft.	131 bags	1.95 lbs. per sq. ft.
R-37	10.0" thick	NET COVERAGE (no joists)	7.83 sq. ft.	128 bags	1.92 lbs. per sq. ft.
		2"x4" joists/16" center	8.22 sq. ft.	122 bags	1.83 lbs. per sq. ft.
		2"x6" joists/16" center	8.41 sq. ft.	119 bags	1.78 lbs. per sq. ft.
R-32	8.7" thick	NET COVERAGE (no joists)	9.00 sq. ft.	111 bags	1.67 lbs. per sq. ft.
		2"x4" joists/16" center	9.50 sq. ft.	105 bags	1.58 lbs. per sq. ft.
		2"x6" joists/16" center	9.76 sq. ft.	102 bags	1.54 lbs. per sq. ft.
R-30	8.0" thick	NET COVERAGE (no joists)	9.78 sq. ft.	102 bags	1.54 lbs. per sq. ft.
		2"x4" joists/16" center	10.39 sq. ft.	96 bags	1.44 lbs. per sq. ft.
		2"x6" joists/16" center	10.70 sq. ft.	93 bags	1.40 lbs. per sq. ft.
R-24	6.5" thick	NET COVERAGE (no joists)	12.04 sq. ft.	83 bags	1.25 lbs. per sq. ft.
		2"x4" joists/16" center	12.95 sq. ft.	77 bags	1.16 lbs. per sq. ft.
		2"x6" joists/16" center	13.41 sq. ft.	75 bags	1.12 lbs. per sq. ft.
R-22	6.0" thick	NET COVERAGE (no joists)	13.04 sq. ft.	77 bags	1.16 lbs. per sq. ft.
		2"x4" joists/16" center	14.13 sq. ft.	71 bags	1.07 lbs. per sq. ft.
		2"x6" joists/16" center	14.66 sq. ft.	68 bags	1.02 lbs. per sq. ft.
R-19	5.2" thick	NET COVERAGE (no joists)	15.05 sq. ft.	66 bags	.99 lbs. per sq. ft.
		2"x4" joists/16" center	16.50 sq. ft.	61 bags	.92 lbs. per sq. ft.
		2"x6" joists/16" center	16.93 sq. ft.	59 bags	.89 lbs. per sq. ft.
R-15	4.0" thick	NET COVERAGE (no joists)	19.57 sq. ft.	51 bags	.77 lbs. per sq. ft.
		2"x4" joists/16" center	22.01 sq. ft.	45 bags	.68 lbs. per sq. ft.
		2"x6" joists/16" center	22.01 sq. ft.	45 bags	.68 lbs. per sq. ft.
R-13	3.5" thick	NET COVERAGE (no joists)	22.36 sq. ft.	45 bags	.68 lbs. per sq. ft.
		2"x4" joists/16" center	25.15 sq. ft.	40 bags	.60 lbs. per sq. ft.
		2"x6" joists/16" center	25.15 sq. ft.	40 bags	.60 lbs. per sq. ft.
R-11	3.0" thick	NET COVERAGE (no joists)	26.09 sq. ft.	38 bags	.57 lbs. per sq. ft.
		2"x4" joists/16" center	29.35 sq. ft.	34 bags	.51 lbs. per sq. ft.
		2"x6" joists/16" center	29.35 sq. ft.	34 bags	.51 lbs. per sq. ft.
BLOWN CAVITY					
R-13	3.5"	NET COVERAGE (no joists)	———	———	———
		2"x4" joists/16" center	16.52 sq. ft.	61 bags	.91 lbs. per sq. ft.
R-20	5.5"	NET COVERAGE (no joists)	———	———	———
		2"x6" joists/16" center	10.52 sq. ft.	95 bags	1.43 lbs. per sq. ft.

TABLE 10.1. INSULATION FACT CHART. (*COURTESY OF PAL-O-PAK*)

Aluminum foil has a perm rating of 0, and if perfectly installed would allow no moisture vapor to pass. Other materials in use have some permeability but are effective retarders.

Moisture vapor is generated in a suprisingly large quantity in homes simply from breathing, certainly from bathing, and continuously throughout the day from clothes washing, dish washing, and food preparation.

Moisture vapor behaves somewhat like heat. It passes through the structure and condenses or forms water drops when it hits a cold surface. To stop this from happening, in walls especially, vapor retarders are installed at the time that insulation is put in place and are usually placed toward the winter-heated space. In fact, all this insulation that we have been talking about can help to create a moisture problem because all cracks, leaks, and the like have been sealed; these aid in carrying away excessive moisture. So, with them sealed, the moisture remains in the home.

In new homes or in homes where installation of a moisture retarder is possible, the material should be installed as follows:

1. All walls and floors exposed to outside temperatures should have a separately applied vapor retarder.
2. The vapor retarder should be located on the warm or heated side of the insulation; this applied to all surfaces.
3. The vapor retarder should be secured to a firm support at all openings, such as windows, doors, electrical outlets, and so on.
4. A vapor retarder should be installed over dirt floors or crawl spaces and basements.
5. A vapor retarder should be installed under all on-grade and below-grade concrete slabs.
6. All joints should have an overlap of at least 3 in.

In existing homes where it is impossible or impractical to install a separate vapor barrier, it is recommended that paint that retards the flow of moisture be used on the inside room surfaces. Reputable dealers can show figures illustrating the relative ability of various brands of paint to resist moisture penetration.

There are also mechanical devices that will help provide the home with moisture control. Exhaust fans, such as the ones normally installed in the bathrooms or kitchen, can be used to control moisture. The fans should be vented to the outside of the building through the roof or wall and be equipped with an automatic damper.

The fans should be controlled by an automatic humidistat connected in parallel with the manual fan switch. When connected in this manner, the user may turn the fan on or off at any time when the humidistat is not calling for the removal of moisture. However, if the humidistat is set for, say, 35 to 40% relative humidity and moisture content becomes greater than this amount, the fan will operate automatically, overriding the manual control.

During summer months, the humidistat should be set at 100% relative humidity or at the OFF position to prevent the fan from running continuously. Of course, a dehumidifier may also be used, but these are often unnecessary.

STARTING THE PROJECT

The following are suggestions made by the manufacturers of various insulating materials:

- Do not cover eave vents or block air passage space along the edge of the roof. Use a baffle if you are pouring in loose fill or if the batt insulation is more than 6 in. thick.

- Do not cover recessed lighting fixtures or exhaust fan motors. Box these off if loose-fill insulation is used.
- Do not overlook any attic areas where there are heated spaces below.
- Push insulation as far as you can under floored areas of the attic.
- Never wear contact lenses when handling insulation.
- Work in the attic in the morning or on a cool cloudy day. Temperatures in the attic can reach 140°F.
- Watch out for nails sticking through roof sheathing or subflooring.
- Take a cold shower when you finish. Cold water closes pores and washes off particles of insulation.

Uninsulated Ceiling: Unroll blankets of insulation, or place batts between attic joists. When you encounter wiring, slip the material under the wires. When bracing is encountered, cut the material and place it tightly above and below it. If a vapor retarder is used, make sure that it is placed downward toward the heated living area of the home.

Start from the edges of the attic and work toward the center. In this way, cutting will probably occur where there is more headroom.

If you have chosen to pour in loose-fill insulation, simply pour the insulation from the bag into the space between joists to the thickness needed to give you the R-value you need. Use a bamboo rake or a board to smooth the insulation to a uniform thickness. A contractor usually installs blown-in loose fill using special pneumatic machinery, with much the same results as when loose fill is poured in.

Vapor retarders need to be added when pouring or blowing in loose fill. Plastic sheeting may be used, or the interior surface of the ceiling may be painted with special vapor-retarding paint or wallpapered with plastic-coated wallpaper.

Partially Insulated Ceiling: Use batts, blankets, or loose fill to add insulation to get the R-values you need. Buy batts and blankets without attached vapor retarders. Otherwise, knife-cut the retardant material about every foot, or tear it off. An added vapor retarder on top of existing insulation would trap moisture vapor.

These types of insulation may be used in any combination. First, determine what you have in the way of R-values and vapor retarders. Then simply add a vapor retarder to the ceilings below if you need one and add insulation to the attic floor to achieve the full R-value level you need.

Insulating Walls: Determine how much insulation you have, if any. To check, turn off the electricity and remove the cover of a convenience outlet or light switch plate located on an outside wall.

If there is some insulation in the wall cavity, do not plan to add more. If there is none, the walls can be insulated by a contractor with special equipment to blow in or foam in place the material you select.

Holes about 2 in. in diameter are drilled in each wall cavity. The insulation is put in through the holes and the holes are then plugged with a precut wooden plug. Where possible, holes are drilled from the outside.

A vapor retarder will have to be added to the inside surface of all outside walls in the form of two coats of oil-based paint made especially for this purpose, or plastic or aluminum foil–coated wallpaper.

The job is simpler and less costly when wall cavities are exposed, as when a house is under construction or complete remodeling is being done. Often studs are exposed in unfinished garages. Blankets and batts with attached vapor retarders are used. The material is stapled in place.

If the vapor retarder is kraft paper, simply staple the flange of the material to the edge or face of the stud facing you inside the house. Two overlapping flanges can be stapled at once, spaced every 3 to 5 in. apart. Staple only the flange. Take care not to allow insulation to lap over on the stud face creating a bulge that will show on the finished wall.

If the vapor retarder is foil, staple the flange to the side of the stud facing you. This creates a 3/4-in. air space between the foil and the finished wall installed later.

If the batts or blankets have no attached vapor retarder, press them into the wall cavity, compacting them as little as possible. Fill cracks around windows and doors and staple 6-mil polyethylene over the entire wall, including windows and doors. When finished, cut the polyethylene away from the openings. Cracks around openings (windows and doors) should be stuffed and covered with polyethylene when using the materials with attached vapor retarders.

Rigid insulating boards achieve higher R-values and are installed on the outer side of the wall cavity in new and full remodeling jobs.

Insulating Ducts: Ducts passing through unconditioned space (not heated or cooled) must be insulated with special insulation manufactured for this purpose. Use the thicker type—2 in. thick with a vapor retarder attached.

Check joints in ducts first and tape them with duct tape if there is any looseness or spaces where ducts might leak. Wrap the ducts with the rolls of insulation, with the vapor retarders outside in this case. Seal the joints formed by wrapping with 2-in.-wide duct tape.

Insulating Floors: The most effective method of insulating floors is by installing batts or blankets between floor joists in unheated crawl spaces and basements. Buy insulation with a vapor retarder, preferably foil in this case, for better insulation of the air space formed. Place the batts, forming the insulation at the girder carefully in a well-fitting manner.

Insulation supporters may be placed between joists every 2 to 3 ft to hold the material in place, but stapling to the bottom of joists does a better job. Work from outside to center, as in the attic. Staple a section of wire to the bottom of joists and slide batts in on top of the wire.

The house is now fully insulated but not yet fully weatherized. Two more steps complete the process: installing storm windows and doors, and weather stripping and caulking.

FINISHING THE PROJECT

Storm Windows: Several methods can be used to form an insulating dead air space of ½- to 4½-in. thickness. A minimum ½ in. of air space between parallel surfaces is required to provide insulating value. Less than ½ in. is acceptable only if air is evacuated from the space, forming a vacuum.

Polyethylene: This type of covering sealed over an existing window is an effective, though short-term solution. Four- to six-mil polyethylene rolls or prepackaged kits with plastic sheets, tacks, strips (and instructions) can be used.

Prefabricated metal and wooden storm windows may be purchased. Metal storm windows usually have two or three tracks, are adjustable for summer ventilation, and have removable glass panels for cleaning.

Wooden storm windows are equally effective and preferred for use in extremely cold climates, since condensation and heat conduction is less with wood frames. Summer storage space is required and frames need repainting periodically.

Insulated glass in permanent windows is an excellent solution most often employed in new construction or remodeling. Two sheets of glass are placed in a single frame and the appearance is similar to single-glazed windows. There are only two surfaces to clean. Breakage replacement costs are doubled, however.

Replacing Windows: Many homeowners are choosing to replace the single-pane windows in their homes with new double- or triple-glazed units. This is a good solution.

Replacement windows come in both wood and metal. Some wooden ones may be covered with vinyl for maintenance purposes. But a metal replacement window should be used only if a thermal break has been built into the frame and insulated or thermal glass is provided. Without a thermal break, heat will be rapidly conducted out of the house by the metal.

Storm Doors: These doors are usually prefabricated metal, although wooden ones are sometimes used with equal effectiveness. Rigid frames, tempered safety glass or rigid nonbreakable plastic, automatic closing devices, and strong safety springs are important considerations. Convertible screen/storm doors are available and popular.

Value will be added to your home with the installation of all types except plastic sheeting. Strength of frames, good quality, warranties, and repair service are important considerations. If you do not do the entire job at one time, do the side of the house facing the prevailing winter winds first, the north side second, and the south side last.

Weather stripping: This comes in several forms and materials. Some are designed for use in one place and some are more durable than others. Table 10–2 covers commonly available types.

Door Bottoms: These are sometimes fitted with as large as a ¼-in. crack left at the bottom. That is equivalent to a 9-in. hole through your wall. Check this and use the best method to seal the crack, depending on the size.

A brass-plated strip fastened to felt or vinyl may be attached to the inside bottom of the door. An even threshold is required and level application is a little tricky to achieve. Other types of sweeps, even automatically operated sweeps, are available.

Thresholds may be improved to seal cracks. Replaceable vinyl bulb-shaped gaskets mounted in metal are effective when properly maintained. Combinations of door bottoms and thresholds are effective and longer wearing.

Caulking: This is the last material to be considered, but it may be the first used in the home. For a relatively small cost and time investment, large savings are produced.

Caulking comes in toothpaste-type tubes and even 5-gallon buckets. But most often a caulking gun with a tube that fits it is best. The tubes have directions for use and suggested places where they work best.

There are a wide variety of products and prices. The less expensive may last only 3 to 5 years, whereas the more expensive may last as long as 30 years.

Something is available for sealing any crack or seam that you find. Sealing exterior spaces reduces the entry of dirt and moisture as well as air. A firm, clean surface is always required for good application. The following are some places that you should check around the home for cracks and seams:

- Joints between door frames and siding
- Joints between windowsills and siding

TABLE 10.2. WEATHERSTRIPPING

Form	Installation/use	Notes
Self-adhesive foam tape	Apply to clean, dry surfaces at room temperature by pressing in place on door and window jambs, stops, or sashes.	Resilient sponge rubber or vinyl on paper or vinyl backing, 3/8 special to 3/4 in. wide; deteriorates when exposed to weather; may last only one season.
Felt or aluminum and felt vinyl	Staple to wood or glue to metal stops, sills, and sashes.	Felt tears easily during use and is ineffective when wet.
	Tack, staple, screw, or glue flange of tube-shaped strip to surfaces.	Durable; easy to apply.
Neoprene-coated sponge rubber	Tack or staple to surfaces.	Easy to install; more durable than uncoated material.
Bronze metal	Tack to door and casement window jambs.	Durable; easy to install; not affected by moisture and temperature.
Caulking cords	Press into place on any type of surface.	Comes in strips; easy to apply; pliable; durable; not affected by moisture.
Fiberglass strip	Various sizes with waterproof tape seal larger cracks, as around garage doors, or may be wrapped around pipes for insulation.	Durable.
Waterproof tape	Seals crack. Apply half on window sash and half on stops. Seals cracks by pressing to clean, dry surfaces.	Not affected by moisture.
Air-conditioner weather strip	Easy to install, rectangular polyfoam strip for sealing around window-mounted units and window sashes.	Low cost.
Magnetic vinyl	For steel door insulation.	Durable.

- Joints between window frame and siding
- Joints between window drip cap and siding
- Inside corners of a house formed by siding
- Joints where two things come together, such as where steps and porches join the main part of the house

- Joints where the chimney and siding come together
- Around chimney and vent pipe flashings
- Places where pipes, wires, or vents pass through exterior walls

Caulk hairline cracks as well as larger ones that you find.

PRIORITIES FOR THE JOB

Complete weatherization of a home may be done gradually rather than all in one big effort. Except for insulating finished walls, this is a do-it-yourself job for a homeowner in most cases.

The groupings below suggest an order to follow. The most effective steps are listed first with the less effective ones toward the end. This list is a general one that does not hold true for each home, since heat loss depends on climate, construction of the house, shape, wall area, window and door area, and so on. But, in general, follow the groupings.

First	Second	Last
Insulate the ceiling Weather-strip and caulk	Install storm windows and doors Insulate walls	Insulate floors

A completely weatherized house will use up to half as much fuel as it uses when not weatherized. These savings come every year and the comfort of residents is increased in the process.

There is a continued heat loss from completely weatherized homes. You have simply cut the rate of loss. Infiltration continues, certainly, as you open and close doors. Management of heated areas, together with the living habits of family residents, are large factors in energy use. The basic fact remains that people—not houses or cars—are the users of energy.

Besides the infiltration of air into homes, windows and other glass areas are the big robbers. You can insulate a wall from a value of R-11 to R-24 with little difficulty. But double glass gives you an R-value of only about 2. Triple glass improves the R-value to something approaching 3. Thus windows are the big concern.

Most heat is lost through north-facing windows. About an equal, though smaller, amount is lost through east and west ones. South-facing windows, carefully managed, may break even or be of advantage in gain and loss measurements.

In new construction and in remodeling, window areas must be carefully weighed for value and function in providing your visual and psychological linkage with the world outside, needed natural ventilation, and continuous heat gain and loss. Placement and size to do the job must be considered carefully and weighed in value systems.

Control and management for best advantage become life-style changes you need to make, perhaps first by temporarily sealing and covering some windows at night or when you are away or in unused rooms. Certainly you can operate thermally effective coverings—such as shades and draperies—as automatically as you brush your teeth each day. An awareness and life-style change will develop. A completely weatherized home is the important first step.

11

HEATING CALCULATIONS

The goals of a heating installation are to obtain adequate, dependable, and trouble-free installations, year-round comfort, reasonable annual operating cost, reasonable installation cost, and systems that are easy to service and maintain.

Heat-loss calculations must be made to ensure that heating equipment of proper capacity will be selected and installed. Heat loss is expressed in either Btu per hour (Btu/h) or in watts. Both are measures of the rate at which heat is transferred and are easily converted from one to the other:

$$\text{watts} = \frac{\text{Btu/h}}{3.4}$$

$$\text{Btu/h} = \text{watts} \times 3.4$$

Basically, calculation of heat loss through walls, roof, ceilings, windows, and floors requires three simple steps:

1. Determine the net area in square feet.
2. Find the proper heat-loss factor from Table 11-1.
3. Multiply the area by the factor; the product will be expressed in Btu/h. Since most electric heat equipment is rated in watts rather than Btu/h, divide this product by 3.4 to convert to watts.

Calculations of heat loss for any building may be made more quickly and efficiently using a prepared form, such as the one shown in Table 11-2.

TABLE 11.1 HEAT-LOSS FACTORS

A. Windows and Doors	
	Watts/ft²
Single-glazed wood sash	21.0
Single-glazed large exposure metal sash	23.1
Single-glazed with storm sash	9.8
Double-glazed ¼ in. air space	11.9
Double-glazed ¾ in. air space	11.0
Solid wood doors, 1⅜ in. thick	10.0
Solid wood doors, 1⅜ in. thick with storm door	8.0
Solid door with air space	12.0

B. Walls	
Frame or 4-in. brick veneer	Watts/ft²
Wood siding, sheathing, plaster, or plasterboard	5.2
With 1 in. insulation	2.7
With 2 in. insulation (R7)	1.8
With 3⅝ in. insulation (R13)	1.2
With 6 in. insulation (R19)	0.8
Frame walls: stucco exterior	
1 in. stucco on metal lath, studs, or plaster	9.2
With building paper and fir sheathing	6.6
With 1 in. insulation	2.9
With 2 in. insulation (R7)	1.8
With 3⅝ in. insulation (R13)	1.2
With 6 in. insulation (R19)	0.9
Solid brick: 8 in. thick	
Only plaster or plasterboard	7.3
With furring and plaster	5.6
With 1⅝ in. insulation	2.5
With 3⅝ in. insulation (R13)	1.2
With 6 in. insulation (R19)	0.9
Concrete block: 8 in. thick	
8 in. block, no finish	10.9
With only plaster or plasterboard	8.8
With furring and plaster	6.4
With 1⅝ in. insulation	2.5
With 3⅝ in. insulation (R13)	1.2
Concrete poured: 8 in. thick	
8 in. concrete, no finish, above grade	14.8
With furring and plaster, above grade	7.4

(*Continued*)

TABLE 11.1 *(Cont.)*

Concrete poured: 8 in. thick	
With 1⅝ in. insulation, above grade	2.7
With 3⅝ in. insulation, above grade	1.4
8 in. concrete, no finish, below grade	2.1
With furring and plaster, below grade	1.8
With 1⅝ in. insulation, below grade	1.2
With 3⅝ in. insulation, below grade (R13)	0.8
Concrete poured: 12 in. thick	
12 in. concrete, no finish, above grade	11.7
With furring and plaster, above grade	6.8
With 1⅝ in. insulation, above grade	2.3
With 3⅝ in. insulation, above grade (R13)	1.2
12 in. concrete, no finish, below grade	1.8
With furring and plaster, below grade	1.6
With 1⅝ in. insulation, below grade	0.8
With 3⅝ in. insulation, below grade (R13)	0.4
Miscellaneous wall construction	
2⅝-in. solid cedar wall	6.2
With 1½-in. fiberglass aluminum-faced batt, furred ½-in. plasterboard	2.1
5-in. solid log wall (no inside finish)	2.9
8-in solid log wall with splined joints (no inside finish)	2.7
Metal-oxidized dull finish, no insulation	2.4
With 1 in. insulation	4.5
With 2 in. insulation (R7)	2.5
With 3 in. insulation (R11)	1.7
With 4 in. insulation (R13)	1.3
With 6 in. insulation (R19)	0.9
Texture one eleven ⅜-in. plywood with air space and ¼-in. plywood	8.0
With 1-in. insulation	3.3
With 2-in. insulation (R7)	2.1
With 3⅝ in. insulation (R13)	1.2
Uni-Panel ¼-in. plywood, 2 in. styrofoam, ¼-in. plywood	1.8

C. Ceilings

Ceilings (below unheated ventilated attic)	Watts/ft^2
Lath and plaster, no flooring above	14.2
Lath and plaster 2 in. insulation and double flooring (R7)	1.8
Lath and plaster 2 in. insulation, no flooring (R7)	2.3
Lath and plaster 4 in. insulation (R13)	1.4
Lath and plaster 6 in. insulation (R19)	1.0
Lath and plaster 8 in. insulation (R26)	0.8
Lath and plaster 10 in. insulation (R33)	0.6
Lath and plaster 12 in. insulation (R38)	0.5

TABLE 11.1 *(Cont.)*

Flat roofs	
2-in. tongue-and-groove built-up roof, no insulation	6.4
2-in. tongue-and-groove built-up roof, 2 in. rigid insulation	2.2
Metal-oxidized dull finish, no insulation	25.9
With 1 in. insulation	4.6
With 2 in. insulation (R7)	2.6
With 3 in. insulation (R11)	1.8
With 4 in. insulation (R13)	1.3
With 6 in. insulation (R19)	0.9
With 8 in. insulation (R26)	0.7
With 10 in. insulation (R33)	0.6
With 12 in. insulation (R38)	0.5
Skylights	
Glass, single sheet	28.8
Glass, double-glazed ¼-in. air space	14.4
Glass, double-glazed over 1-in. air space	12.9
Fiberglass, single corrugated sheet	28.8
Fiberglass, outside, air space, and glass or fiberglass under	12.9
D. Floors	
Ventilated crawl space under floors	Watts/ft^2
Double wood floor, no insulation	5.7
2-in. tongue-and-groove with carpeting	4.3
Double wood floor, 1 in. insulation	3.0
Double wood floor 2 in. insulation (R7)	1.8
Double wood floor, 4 in. insulation (R13)	1.0
Double wood floor, 6 in. insulation (R19)	0.8
Slab floors per linear foot of exposed edge or perimeter	
Slab floor, no insulation, laid over cinder or rock	16.8
Slab floor, 2 in. edge insulation between slab and footing and extending 2 ft. under the perimeter of slab	11.3
Basement floor per square foot	
Concrete floors	0.5

With spaces for all necessary data and calculations, the procedure becomes routine and simple.

The load estimate is based on design conditions inside the building and outside in the atmosphere surrounding the building. Outside design conditions are the maximum extremes of temperature occurring in a specific locality (see Table 11-3). The inside design condition is the degree of temperature and humidity that will give optimum comfort.

TABLE 11.2 HEAT-LOSS CALCULATION FORM

1 and 2	For		Job			Type of Construction
	Design temp.	Heat		Insulation type:		
	Inside temp.	°F		Thickness or R-value for:		First-floor crawl space___ basement___ slab___
2 and 3	Outside temp.	°F		Wall	Floor	Basement heated___ unheated___
	Temp. diff.	°F		Ceiling	Other	Basement wall above grade___ below___
4	Room name					
5	Room dimension	W	L	H		
6	Room size, square feet ($W \times L$)					
7	Exposed wall, linear feet					
8	Gross exposed wall, square feet					

	Source of loss	Watts factor	Area	Watts loss
9	Glass area, square feet			
10	Door area, square feet			
11	Net wall area, square feet			
12	Ceiling area, square feet			
13	Floor area, square feet			
14	Slab edge, linear feet			
15	Room volume, cubic feet (infiltration)			
16	Room totals, watts at 70°F			
17	Adjusted watts (multiplier)			
18	Equipment			
19	Control			
20	Installed watts or btu/h			

TABLE 11.3 HIGH AND LOW AVERAGE TEMPERATURES AND AVERAGE DEGREE-DAYS FOR VARIOUS U.S. CITIES

State/City	Low temp.	High temp.	Degree-days
Alabama			
Anniston	5	96	2820
Birmingham	10	97	2780
Mobile	15	95	1612
Montgomery	10	98	2137
Arizona			
Flagstaff	−10	84	7525
Phoenix	25	108	1698
Yuma	30	67	951
Arkansas			
Bentonville	−5	97	4036
Fort Smith	10	101	3188
Little Rock	5	99	2982
California			
Eureka	30	67	4832
Fresno	25	101	2532
Los Angeles	35	94	2015
Sacramento	30	100	2822
San Diego	35	86	1574
San Francisco	25	80	3421
San Jose	35	90	2410
Colorado			
Denver	−10	90	6132
Grand Junction	−15	96	5796
Pueblo	−20	96	5709
Connecticut			
Hartford	0	90	6139
New Haven	0	88	6026
District of Columbia			
Washington	0	94	4333
Florida			
Apalachicola	25	92	1307
Jacksonville	25	96	1243
Key West	35	90	89
Miami	35	92	178
Pensacola	20	92	1435
Tampa	30	92	674
Georgia			
Atlanta	10	95	2826
Augusta	10	98	2138
Macon	15	98	2049
Savannah	20	96	1710
Idaho			
Boise	−10	96	5890
Lewiston	−15	98	5483
Pocatello	−5	94	6976

TABLE 11.3 *(Cont.)*

State/City	Low temp.	High temp.	Degree-day
Illinois			
Cairo	0	97	3756
Chicago	−10	95	6310
Peoria	−10	94	6087
Springfield	−10	95	5693
Indiana			
Evansville	−10	96	4360
Fort Wayne	−10	93	6287
Indianapolis	−10	93	5611
Iowa			
Charles City	−25	91	7504
Davenport	−15	94	6091
Des Moines	−15	95	6446
Dubuque	−20	92	7271
Sioux City	−20	96	7012
Kansas			
Concordia	−10	101	5323
Dodge City	−10	99	5058
Topeka	−10	99	5209
Wichita	−10	102	4571
Kentucky			
Lexington	−5	94	4979
Louisville	0	96	4439
Louisiana			
New Orleans	20	93	1317
Shreveport	20	99	2117
Maine			
Eastport	−10	85	8246
Portland	−5	88	7681
Maryland			
Baltimore	0	94	4787
Massachusetts			
Boston	0	91	5791
Michigan			
Detroit	−10	92	6404
Escanaba	−20	82	8657
Grand Rapids	−10	91	7075
Houghton	−20	—	9030
Lansing	−10	89	6982
Ludington	−10	87	7458
Marquette	−20	88	8529
Sault Ste. Marie	−20	83	9475
Minnesota			
Duluth	−25	85	9937
Minneapolis	−20	92	7853

(Continued)

TABLE 11.3 *(Cont.)*

State/City	Low temp.	High temp.	Degree-days
Moorehead	−30	92	9327
St. Paul	−20	92	7804
Mississippi			
Corinth	0	98	3087
Meridian	10	97	2333
Vicksburg	10	97	2000
Missouri			
Columbia	−10	97	5113
Hannibal	−15	96	5393
Kansas City	−10	100	4888
St. Louis	0	98	4699
Springfield	−10	97	4693
Montana			
Billings	−35	94	7106
Havre	−30	91	8213
Helena	−20	90	8250
Kalispell	−35	88	8055
Miles City	−35	97	7850
Missoula	−20	92	7873
Nebraska			
Lincoln	−10	100	6104
North Platte	−20	97	6546
Omaha	−20	97	6160
Valentine	−25	97	7075
Nevada			
Reno	−5	92	6036
Tonopah	−10	92	5813
Winnemucca	−15	95	6369
New Hampshire			
Concord	−15	88	7612
New Jersey			
Atlantic City	5	91	4741
Cape May	—	91	4870
Newark	0	94	5252
Sandy Hook	0	—	5369
Trenton	0	92	5068
New Mexico			
Albuquerque	0	96	4389
Roswell	0	99	3424
Santa Fe	0	88	6123
New York			
Albany	−10	88	6962
Binghamton	−10	80	7537
Buffalo	−5	86	6838
Canton	−25	86	8305
Ithaca	−15	91	6914
New York	0	94	5050

TABLE 11.3 (*Cont.*)

State/City	Low temp.	High temp.	Degree-days
Oswego	−10	86	6975
Rochester	−5	91	6836
Syracuse	−10	89	6520
North Carolina			
Asheville	0	91	4072
Charlotte	10	96	3205
Hatteras	20	—	2392
Raleigh	10	95	3369
Wilmington	15	93	2323
North Dakota			
Bismark	−30	95	9033
Devils Lake	−30	93	9940
Grand Forks	−35	91	9871
Williston	−35	94	9068
Ohio			
Cincinnati	0	94	5195
Cleveland	0	91	6006
Columbus	−10	92	5615
Dayton	0	92	5597
Sandusky	0	92	5859
Toledo	−10	92	6394
Oklahoma			
Oklahoma City	0	100	3519
Oregon			
Baker	−5	94	7087
Medford	5	98	4547
Portland	10	89	4632
Rosebug	10	93	4122
Pennsylvania			
Erie	−5	88	6116
Harrisburg	0	92	5258
Philadelphia	0	93	4866
Pittsburgh	0	90	5905
Reading	0	92	5060
Scranton	−5	89	6047
Rhode Island			
Block Island	0	—	5843
Providence	0	89	6047
South Carolina			
Charleston	15	95	1973
Columbia	10	98	2435
Greenville	10	95	3060
South Dakota			
Huron	−20	97	7902
Pierre	−25	98	7283
Rapid City	−20	96	7535

(*Continued*)

TABLE 11.3 *(Cont.)*

State/City	Low temp.	High temp.	Degree-days
Tennessee			
Chattanooga	10	97	3384
Knoxville	0	95	3590
Memphis	0	98	3137
Nashville	0	97	3513
Texas			
Abilene	15	101	2657
Amarillo	−10	98	4345
Brownsville	30	94	617
Corpus Christi	20	95	1011
Dallas	0	101	2275
El Paso	20	100	2641
Ft. Worth	10	102	2361
Galveston	20	91	1233
Houston	20	96	1388
Palestine	15	99	1980
Port Arthur	20	94	1517
San Antonio	20	99	1579
Taylor	10	101	1909
Utah			
Modena	−15	—	6598
Salt Lake City	−10	97	5866
Vermont			
Burlington	−10	88	7865
Northfield	−20	86	8804
Virginia			
Cape Henry	10	—	3307
Lynchburg	5	94	4153
Norfolk	15	94	3454
Richmond	15	96	3955
Wytheville	0	92	5103
Washington			
North Head	20	—	5211
Seattle	15	94	4153
Seattle–Tacoma	10	85	5275
Spokane	−15	93	5862
Tacoma	15	85	4866
Tatoosh Island	15	—	5724
Walla Walla	−15	98	4848
Yakima	−5	94	5845
West Virginia			
Elkins	−10	87	5773
Parkersburg	−10	93	4750
Wisconsin			
Green Bay	−20	88	8259
LaCrosse	−25	90	7650

TABLE 11.3 *(Cont.)*

State/City	Low temp.	High temp.	Degree-days
Madison	−15	92	7417
Milwaukee	−15	90	7205
Wyoming			
Cheyenne	−15	89	7562
Lander	−20	92	8303
Yellowstone Park	−35	90	9605

To use the form in Table 11-2, proceed with the following steps. The step number corresponds with the space numbers on the form.

1. Enter for whom calculation is being made, job title, and location.
2. Establish all necessary construction and insulation characteristics and insert in appropriate spaces.
3. Insert local design temperature for the area in which the building is located. Determine the required temperature change to obtain a 70°F ambient inside temperature.
4. Insert individual room names.
5. Measure individual rooms and insert overall length (L), width (W), and height (H) dimensions.
6. Determine room size ($W \times L$).
7. Determine linear feet of exposed outside wall.
8. Determine gross exposed wall (No. 7 $\times$ H).
9. Measure individual windows for each room, insert values in square feet, then multiply times the factor (Table 11-1A).
10. Measure individual outside doors, insert values in square feet, then multiply times the factor (Table 11-1A).
11. Subtract the total areas (square feet) of lines 9 and 10 from line 8 and insert value of net exposed wall, then multiply times the factor (Table 11-1B).
12. Insert ceiling area square feet (same as room size, line 6), then multiply times the factor (Table 11-1C).
13. Insert floor area square feet (same as room size, line 6), then multiply times the factor (Table 11-1D) *or* use line 14 for slab floors.
14. Insert slab edge, linear feet (same as line 7), then multiply times the factor (Table 11-1D).
15. Insert room volume, cubic feet ($W \times L \times H$), then multiply by 0.3 (standard for ¾ air change per hour) or 0.04 (standard for basements below grade).

TABLE 11.4 OUTSIDE DESIGN TEMPERATURE CORRECTION MUTLIPLIER FOR INDOOR DESIGN TEMPERATURE OF 70°F

If outside design temperature (°F) in your area is:	Multiply watts required by:
0	1.00
+5	0.93
+10	0.86
+15	0.79
+20	0.72
+25	0.65
+30	0.57
+35	0.51
+40	0.44
−5	1.07
−10	1.15
−15	1.22
−20	1.29
−25	1.36
−30	1.43
−35	1.50

16. For total watts, total the "watts loss" column (lines 9 through 15).
17. For "adjusted watts," determine (and insert) multiplier from Table 11-4, then multiply times line 16 for each room and insert (if design temperature is 0, lines 16 and 17 would be the same).
18. Insert catalog numbers for the heater(s) required to approximate most closely the "adjusted watts" (line 17) figure for each room.
19. Insert the catalog number for the controls for each room.
20. Insert the total installed wattage (or Btu/h) for each room.

AIR CONDITIONING

Although the installation of air-conditioning systems usually falls under the mechanical section of a building's systems, electricians are required to furnish power—and sometimes do the control wiring—for such systems. On smaller projects, such as residential construction and small commercial projects, the electrical contractor may be called upon to furnish and install such items as ventilating fans, through-wall room air conditioners, duct heaters, and the like. For these reasons, anyone involved in electrical technology should have a basic understanding of air-conditioning systems.

12

HIGH-VELOCITY HVAC SYSTEMS

Today, as never before, many existing structures are purchased and renovated rather than building a new home. In nearly all of these cases, many major repairs are necessary to make the old buildings safe and convenient to live in. Of these major repairs, the heating and cooling system is usually the most expensive due to the cutting and patching of the existing structure to fit in the new system. When it becomes necessary to run several large air ducts, this cost can be tremendous. However, much of this expense can be saved by installing a high-velocity heating and cooling system.

While there are several high-velocity HVAC systems on the market, Dunham-Bush, Inc. of Harrisonburg, Virginia, manufactures Space-Pak kits especially for residential use and many have been completely installed by homeowners themselves.

THE SYSTEM COMPONENTS

The basic components of a residential high-velocity heating and cooling system are shown in Fig. 12-1. In general, the system consists of the following individual items:

1. Return air assembly
2. Blower coil unit
3. Condensing unit
4. Duct heater
5. Plenum duct
6. Air-supply ducts
7. Air-supply outlet plate

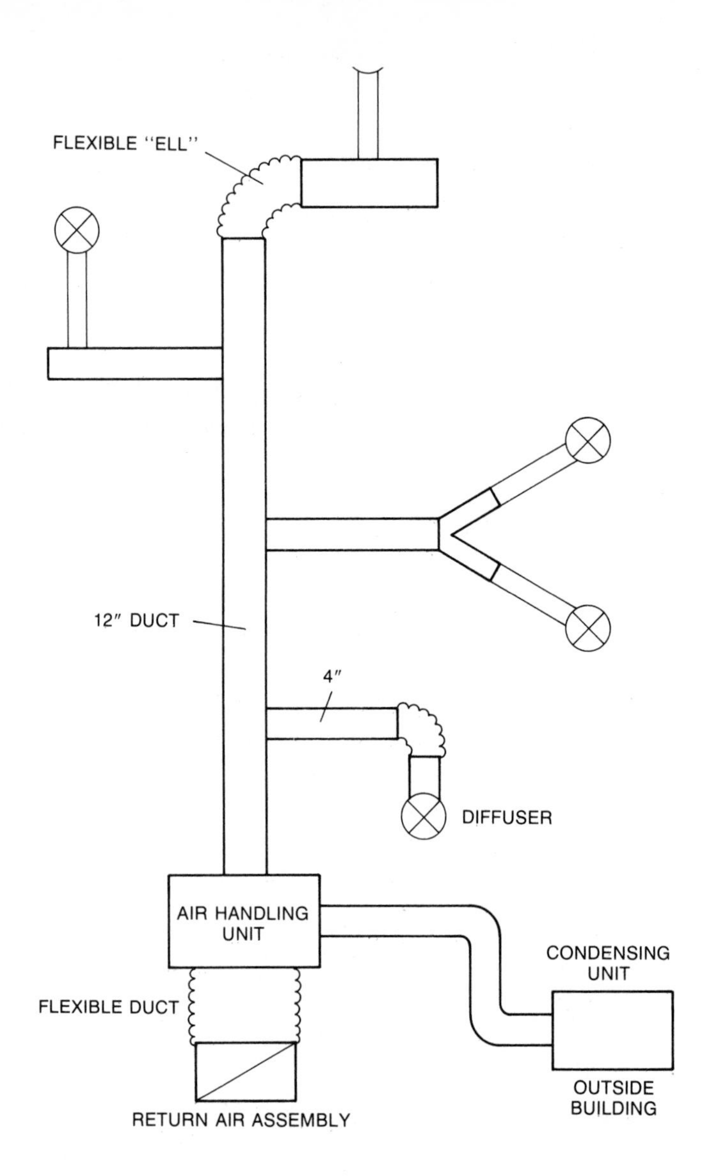

Figure 12–1. Basic components of a high-velocity residential heating and cooling system

Return-Air Assembly: This return air assembly should be located in a central area of the home, and its opening probably requires the only major cutting in the entire system. An optional air cleaner is available which electrostatically removes dust particles from the air to keep the home cleaner.

Blower-Coil Unit: Attic or basement installation of a high-velocity blower unit is compact and neat. Most kits include a spring-mounted, rubber-isolated blower and electric motor plus an expansion valve, condensate overflow control, and an extra-deep cooling coil that removes up to perhaps 40%

more humidity from the air than shallow ones. Sizes are available from 17,000 to over 50,000 Btu/h. A prefabricated 7-foot length of 14- or 18-in. flexible return-air duct connects the return assembly to the blower unit.

Condensing Unit: The condensing unit is the only piece of outside equipment needed. The twin blowers housed in an aluminum cabinet can be mounted by several different methods, some of which are described as follows:

1. Give much thought to the location and mounting of the condensing unit, preferably before the components are ordered; then the exact length of refrigerent tubing can be correctly ordered.
2. The condensing unit should be mounted outside the building, as close as possible to the cooling coil inside the air plenum. Furthermore, it should be installed on a solid, level support.
3. For roof installations, wooden beams (treated to reduce deterioration) can be used, as can channel-iron supports. Both should be sized to support the weight of the condensing unit and extended beyond the unit to distribute the load on the roof. Any local iron works can make a support frame once they have the exact dimensions.
4. For ground installations, use piers or a concrete slab with footers extended below the frost line, wooden beams treated to reduce deterioration, or a channel-iron frame with a suitable base.
5. In some situations, it may be necessary to mount the condensing unit on the side or wall of a building. If so, use angle-iron mounting frames dimensioned to fit the condensing unit and supported adequately to hold the weight of the unit.
6. If it becomes necessary to mount the unit more than about 2 ft from the building, provisions must be made to run both the electrical and refrigeration lines. An electrical conduit will be required for the power supply and a separate conduit for the low-voltage control wiring. However, if the unit is mounted within 2 ft of the building, the fused disconnect switch can be mounted on the side of the building and the wiring and refrigerant lines can loop directly from the side of the house to the condensing unit.
7. Adequate clearance must be provided for: air intake and air discharge, refrigerant piping and power connections, and maintenance and servicing. The recommended minimum clearances can vary from 6 in. to 6 ft.

Duct Heater: The optional electric heating system can be added in increments from 5 to 25 kilowatts or 17,000 to 85,000 Btu/h. The heating system is installed in the main plenum duct just downstream of the blower coil.

Plenum Duct: Seven-inch sound-absorbing fiberglass duct is used for the plenum. An aluminum vapor barrier on the outside of the plenum controls moisture; the flexible 2-in. supply ducts are easily cut into the supply plenum.

Air-Supply Ducts: Preinsulated flexible supply ducts quickly connect to the main plenum duct; two parts join with a twist of the wrist, giving an airtight seal for good air delivery. The small (3½-in.) ducts are easily run between studs and around obstacles, requiring less cutting and patching than conventional forced-air systems. An assortment of factory-made fittings helps speed the installation.

Air-Supply Outlet Plate: The tiny 2-in. opening will heat and cool the average room; it can be located in an out-of-the-way corner in ceiling or floor so that it is hardly noticeable. Adjustable dampers built into each outlet permit you to control the airflow according to the season or the number of people in the area.

Controls: Besides a conventional thermostat and control panel, the controls include a factory-installed antifrost control which automatically shuts off the compressor if icing of the cooling coil occurs, and a mercury float switch to prevent condensate overflow.

SYSTEM DESIGN

As with any system, proper sizing of a high-velocity system requires accurate calculation of heat gain and loss (see Chapter 11).

System layout is a different story; only personnel trained in the design and application of high-velocity systems should attempt to lay one out. The calculations required are complicated and critical; the numerous design considerations include the thermal load on, and the length of duct run to, each area.

When such a system is to be installed by other than a factory-trained contractor, leave the design to the manufacturer's engineer, furnishing him or her with a complete set of working drawings for the building in question. The engineer will then lay out the system completely, locating duct runs, return and supply fixtures, and all other system hardware at little or no cost to the homeowner. In addition, the engineer will probably give the homeowner hints on installing the system and alert you to the most common mistakes.

Once the system has been designed to suit a particular building, the small preinsulated ducts and easily connected components make it possible to install the entire system in a short period of time.

TROUBLESHOOTING HIGH-VELOCITY SYSTEMS

Troubleshooting high-velocity HVAC systems covers a wide range of electrical and mechanical problems, from finding a short circuit in the power supply line, through adjusting a pulley on a motor shaft, to tracing loose connections in complex control circuits. However, in nearly all cases, the troubles can be determined by using a systematic approach, checking one part of the system at a time in the right order.

The following data are arranged so that the problem is listed first. Then the possible causes of this problem are listed in the order in which they should be checked. Finally, solutions to the various problems are given, including step-by-step procedures where it is believed that they are necessary.

Since troubleshooting electrical resistance problems was covered earlier in this book, only the troubleshooting methods for the cooling mode are covered in this section.

Problem 1: *The compressor motor and condenser motor will not start, but the fan/coil unit operates normally.*

Remedy

1. Check the thermostat system switch to ascertain that it is set to COOL.
2. Check the thermostat to make sure that it is set below room temperature.
3. Check the thermostat to see if it is level. Most thermostats must be mounted level; any deviation will ruin their calibration. To correct, remove the cover plate, place a spirit level on top of the thermostat base, loosen the mounting screws, and adjust the base unit until it is level; then tighten the mounting screws.
4. Check all low-voltage connections for tightness.
5. Make a low-voltage check with a voltmeter on the condensate float switch: the condensate may not be draining. The float switch is normally found in the fan/coil unit. A typical switch consists of a 24-V, normally closed mercury switch attached to a stainless-steel arm and pin with a polystyrene float. The switch opens if the water level in the drain pan rises to approximately ½ in. The water level in the drain pan should be between ¼ and ⅜ in. when the unit is running.

   To make the low-voltage check, set the thermostat on cooling function and, with the fan running, check across the two lead wires at an accessible point (such as where they connect to a low-voltage terminal block). You should get a reading of 24 V. If the float switch is broken, the mercury bulb is loose, the contact does not open or close, or the float is loose, replace the float switch assembly.

   If you find the float switch to be in good working order, clean out the condensate drain and trap; then check to see if the fan/coil unit is level. Continue by checking the pitch (¼ in. for each foot of horizontal line) of the condensate drain line.
6. Low airflow could be causing the trouble, so check the air filters; if dirty, clean or replace.
7. Make a low-voltage check of the antifrost control; replace if defective.
8. Check all duct connections to the fan/coil unit; repair if necessary.

Problem 2: *The compressor, condenser, and fan/coil unit motors will not start.*

Remedy

1. Check the thermostat system switch setting to ascertain that it is set to COOL.
2. Check the thermostat setting to make sure that it is below room temperature.

3. Check the thermostat to make sure that it is level. If not, correct.
4. Check all low-voltage connections for tightness.
5. Check for a blown fuse or tripped circuit breaker. Determine the cause of the open circuit and replace the fuses or reset the circuit breaker.
6. Make a voltage check of low-voltage transformer; replace if defective.
7. Check your electrical service against minimum requirements, that is, for correct voltage, ampere rating, wire size, number of circuits, and so on. Update your service if necessary.

Problem 3: *The condensing unit cycles too frequently, the contactor opens and closes on each cycle, and the blower motor operates.*

Remedy

1. The condensate drain may not be working properly. Run a check as described in item 5 of Problem 1. Then continue on to items 6, 7, and 8 of the same problem.
2. Check all low-voltage wiring connections for tightness, and correct if necessary.
3. The blower motor could be defective, so take an amperage reading on the motor while it is running. However, do not confuse the full-load (starting) amperes shown on the motor rating plate with the actual running amperes. The latter should be about 25% or less. If the amperage varies considerably from that on the nameplate, have the motor checked for bad bearings, defective winding insulation, and so on. The problem could also be caused by low supply voltage.

Problem 4: *Cooling is inadequate with the condensing unit and the blower running continuously.*

Remedy

1. Check all low-voltage connections (control wiring) against the wiring diagram furnished with the system. Correct if necessary. Then check for leaks in the refrigerant lines.
2. Check all joints in the supply and return ductwork; make all joints tight.
3. The equipment could be undersized. Check the heat gain calculations against the output of your unit. Correct structural deficiencies with insulation, awnings, and so on, or install properly sized equipment.

Problem 5: *The condensing unit cycles but the blower motor does not run.*

Remedy

1. Check all low-voltage connections against the wiring diagram furnished with the system. Correct if necessary.
2. Check all low-voltage connections for tightness.
3. Make a voltage check on the blower relay and replace if necessary.
4. Make electrical and mechanical checks on the blower motor. Check for correct voltage at motor terminals. Mechanical problems could be bad bearings or a loose blower wheel. Bearing trouble can be detected by turning the blower wheel

by hand (with current off), and checking for excessive wear, roughness, or seizure.

Problem 6: *The blower coil unit short-cycles continuously and cooling is insufficient.*

Remedy

1. Make electrical and mechanical checks as described in item 4 of Problem 5. Repair or replace motor if necessary.

Problem 7: *Sweating appears at the blower coil output or at the electric duct heater outlet.*

Remedy

1. Check to see that the insulation is installed properly.
2. Inspect the joints at the duct heater or blower coil receiving collar; seal properly.

TYPICAL HEATING PROBLEMS

This section details diagnosis and repair procedures for problems specific to the heating function.

Problem 8: *The thermostat calls for heat but the blower motor will not operate.*

Remedy

1. Check all low-voltage connections against the wiring diagram furnished with the system. Correct if necessary.
2. Check all low-voltage connections for tightness.
3. Check line voltage against the unit's nameplate. Correct if necessary.
4. Check all line-voltage connections for tightness.
5. Check for blown fuses or a tripped circuit breaker in the line; determine the reason for the open circuit and replace the fuses or reset the circuit breaker.
6. Check the low-voltage transformer; replace if defective.
7. Make a low-voltage check on the magnetic relay; repair or replace if necessary.
8. Make electrical and mechanical checks on the blower motor as described in item 4 of Problem 5. Repair or replace the motor if defective.

Problem 9: *When the thermostat calls for heat, the blower motor operates but delivers cold air.*

Remedy

1. Make a visual and electrical check on the heating elements. Replace if defective.
2. Make an electrical check on the heater limit switch—first disconnecting all power to the unit—using an ohmmeter to check continuity between the two terminals of the switch. If the limit switch is open, replace it with one of the same rating.

3. Make an electrical check on the time-delay relay. Most are rated at 24 V and have one set of normally open main contacts for line duty and one set of normally open auxiliary contacts for pilot duty. First check the voltage between terminals H^1 and H^2 with the relay heater energized. The reading should be 24 V. Next connect the voltmeter across terminals M^2 on the time-delay relay and L^2 on the terminal block. After a delay of about 45 seconds there should be full line voltage across these contacts and the heating element should be energized. Again, use the voltmeter and check the voltage across terminals A^2 and H^2 on the time-delay relay. This should read 24 V and will indicate if the pilot duty contacts have closed with the line duty contacts. If they do not close after the relay heater has been energized, disconnect the power to the unit and check for continuity between terminals H^1 and H^2 with an ohmmeter. Then check for continuity between each terminal and ground. If the relay heater coil is open or grounded, replace the time-delay relay. If the time-delay relay heater checks and the main and auxiliary contacts do not close, this indicates that the cross bar is binding or misaligned, and the time delay should be replaced.
4. Make an electrical check on the magnetic relay and repair or replace if defective.
5. Check your electrical service entrance and related circuits against the minimum recommendations.

Problem 10: *When the thermostat calls for heat, the blower motor operates continuously and the system delivers warm air, but the thermostat is not satisfied.*

Remedy

1. Check all joints in the ductwork for air leaks, making all defective joints tight.
2. Check all duct joints and blower outlets for tightness and seal where necessary.
3. Make a visual and electrical check of the electric heating element and repair or replace if necessary.
4. Make an electrical check on the heater limit switch as described in item 2 of Problem 9.
5. Make an electrical check on the heater limit switch as described in item 3 of Problem 9.
6. Check the heating element against the blower unit for the possibility of a mismatch.
7. Check your heat-loss calculations. The equipment could be undersized. If so, correct structural deficiencies by installing more insulation, storm windows and doors, and so on, or install properly sized equipment.

Problem 11: *The blower unit operates properly and delivers air but the thermostat is not satisfied.*

Remedy

1. Check all joints in the ductwork for air leaks and repair if necessary.
2. Check the air filter and clean or replace if necessary. Also check the number of air outlets for adequacy and make sure that they are balanced.
3. Check for undersized equipment as described in item 7 of Problem 10.

Problem 12: *The electric heater cycles on the limit switches but the blower motor does not operate.*

Remedy

1. Make an electrical check on the magnetic relay and repair or replace if defective.
2. Make electrical and mechanical checks on the blower motor as described in item 4 of Problem 5.
3. Check the line connection against the wiring diagram furnished with the system.

TYPICAL HEATING AND COOLING PROBLEMS

This section gives troubleshooting and repair procedures for problems affecting both system functions.

Problem 13: *Air makes excessive noise at the terminator.*

Remedy

1. Duct or outlet undersized; air velocity too great. Increase the size of the duct and/or outlet.
2. Make an external static pressure check and correct restrictions in the system if necessary.
3. Check for a properly balanced system and make corrections if necessary.

Problem 14: *The noise is excessive at the return air grille.*

Remedy

1. Check the return duct to make sure that it has a 90° bend in it.
2. Make a visual check of the blower unit to ascertain that all shipping blocks and angles have been removed.
3. Check the blower motor assembly suspension and fasteners and tighten if necessary.

Problem 15: *The vibration at the blower unit is excessive.*

Remedy

1. Visually check for vibration isolators (which isolate the blower coil from the structure). If missing, install as recommended by the manufacturer.
2. Visually check to ascertain that shipping blocks and angles have been removed from the blower unit.
3. Check the blower motor assembly suspension and fasteners and tighten if necessary.

MAINTENANCE OF HEATING AND COOLING EQUIPMENT

The old saying, "an ounce of prevention," certainly holds true for heating and cooling systems. A correctly installed system that is maintained according to the manufacturer's recommendations will give you years of trouble-free service at minimum cost.

Maintenance procedures and frequency will depend on the type of system you have, but the information given here covers most residential heating and cooling systems in general. For further information, consult the owner's handbook accompanying your equipment. If you do not have one, ask your supplier or write directly to the manufacturer to obtain a copy.

Besides saving on repairs, good maintenance of your heating and cooling equipment will ensure that your equipment operates at maximum efficiency, which will save fuel and reduce other operating expenses.

Cleaning: Heating and cooling equipment should be cleaned at regular intervals to maintain operating efficiency, lengthen the life of the equipment, and minimize energy consumption and operating costs.

Begin by removing lint and dust with a cloth and brush from all finned convector-type heaters, and then vacuum them. This includes electric and hot-water baseboard heaters, wall, and unit heaters. Air-conditioning evaporator and condensor coils should be vacuumed, scrubbed with a liquid solvent or detergent, and flushed.

Air filters in forced-air systems collect dust, which is their purpose. Periodic inspection will tell you how often they should be cleaned or replaced. Permanent metal mesh and electrostatic air filters should be washed and treated; throwaway filters should be replaced with the same size and type. Obtain them from your local hardware store or mechanical contractor.

Vacuum or wipe off lint and dust from all supply and return grilles, diffusers, and registers, using a detergent solution if necessary. While you are doing this, also remove any dirt on the dampers and check the levers for proper operation.

Motors should also be vacuumed to remove dust and lint, but you should first turn off the power supply. Once free of dust and lint, wipe their exterior surfaces clean with a rag and reconnect the power.

Propeller fans and blower wheels are especially susceptible to dust deposits and should be cleaned often. Again, make certain that the power is turned off—to prevent losing a finger—and remove the dirt deposits with a liquid solvent.

Every periodic inspection of your heating and cooling system should include a check of the condensate drain. Wash the drain pan with a mild detergent and flush out the drain line. All loose particles of dirt should be brushed from the evaporator coil and a fin comb should be used to open all clogged air passages in the coil. If the coil is extremely dirty, a small pressurized sprayer may be used with a strong dishwasher detergent to flush the coil. Always rinse with clean water after using detergent.

Lubrication: Adequate, regular lubrication ensures efficient operation, long equipment life, and minimum maintenance cost, but never overlubricate. Observe the manufacturer's instructions when lubricating all bearings, rotary seals, and moveable linkages of

1. *Motors:* direct or belt-drive type
2. *Shafts:* fans, blower wheel, and damper
3. *Pumps:* water circulating
4. *Motor controllers:* sequential and damper operators

Periodic Inspection: This checklist should be followed during periodic inspections of your heating and cooling system in order to minimize maintenance expense and to save as much fuel as possible.

1. *Drive belts:* Examine for
   - Proper tension and alignment
   - Sidewall wear
   - Deterioration, cracks
   - Greasy surfaces
   - Safety guards in position and secure
2. *V-pulleys:* Examine for
   - Alignment
   - Wear of V-wall
   - Tightness of pulley and setscrews
3. *Fan blade and blower wheel assemblies:* Examine for
   - Metal fatigue cracks
   - Tightness of hubs and setscrews
   - Balance
   - Safety guards in position and secure
4. *Electrical components*
   - Examine, burnish, or replace electrical contacts of
     Magnetic contactors and relays
     Thermal relays
     Control switches, thermostats, timers, etc.

- Examine wire and terminals for
  Corrosion and looseness at switches, relays, thermostats, controllers, fuse clips, capacitors, etc.
- Examine motor capacitors for
  Case swelling
  Electrolyte leakage

When checking the electrical system, examine all components for evidence of overheating and insulation deterioration.

13

CLEAN AIR

For cleaner air in residential heating applications, many engineers are specifying that ultraviolet sterile conditioners be installed in the duct systems. Although several such systems are available on the market, the American Ultraviolet Co. manufactures one that consists of a compact ultraviolet lamp, transformer, and accessory assembly—providing a complete packaged unit ready to meet practically any residential and commercial application. The residential (D-10) series equipment is available in one- to four-lamp units.

American Ultraviolet Sterile conditioners are adaptable to practically any heating and cooling system, as well as exhaust systems, and installation may be made during or after the ductwork has been installed. Installation is accomplished by cutting a small hole in the duct wall for lamp insertion and fastening the transformer section to the exterior wall surface. For large plenum chambers, a simple support bracket is added for the sterile conditioner. In general, the unit may be used in residential window or larger air-conditioning units.

To the five important functions already performed by a residential air conditioning system—control of air temperature, humidity, circulation, ventilation, and cleanliness—sterile conditioners add one more: air sanitization.

Filters are an effective means of removing dust, pollen, and other large airborne particles. Electronic air cleaners are even better. However, filters alone cannot remove microorganisms, bacteria, and viruses; even the most efficient air filters are only partially effective. Bacteria removal is a job specifically for the germ-killing ultraviolet rays produced by sterile conditioning units.

Wherever air is recirculated—summer and winter, in comfort or process air conditioning—there is a real need for germicidal ultraviolet if complete air sanitization is desired. To illustrate, whenever air is recirculated through any occupied area, the bacteria count rises steadily, and occupants can be affected by this unsanitary air. A sterile conditioner, properly installed in the air system, can sanitize all the air recirculated.

Wherever people congregate, sterile conditioners play an important part in preventing the spread of airborne infections. Tests have revealed that the disinfected air in sterile conditioned systems can have a bacteria count lower than that of the outdoor air, despite the fact that the indoor air is recirculated many times.

INSTALLATION

In the average installation, a 90% kill of airborne microorganisms is recommended. This rate of kill is used most to protect personnel from cross-infection. However, a higher kill rate, at least 98%, is more desirable in areas where absolute sterility is often necessary. The table in Fig. 13-1 indicates the number of lamps required to increase the kill rate to the desired level. This table represents comprehensive research and testing resulting in the exact amount of germicidal ultraviolet energy necessary for various degrees of air disinfection.

Sterile conditioners are available with one to four lamps. This provides installation flexibility so that the correct number of lamps for the desired kill rate can be installed in any air system.

Sterile conditioners are designed for full-efficiency utilization of ultraviolet lamps. When installed, the lamps will be at right angles to the flow of

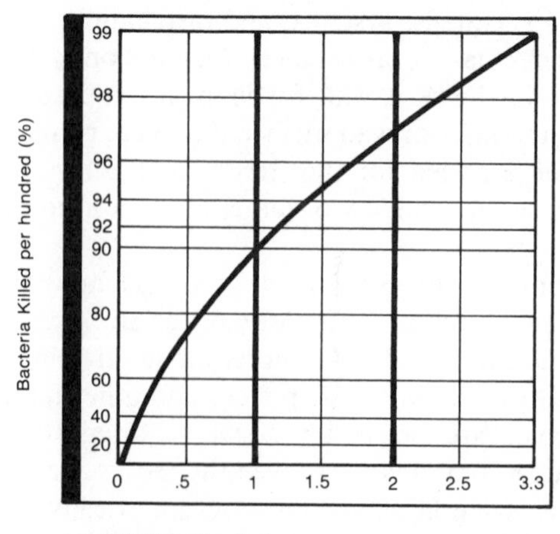

Figure 13-1. Lamp multiplication factors for increasing the kill of bacteria. (*Courtesy American Ultraviolet Company, Chatham, New Jersey*)

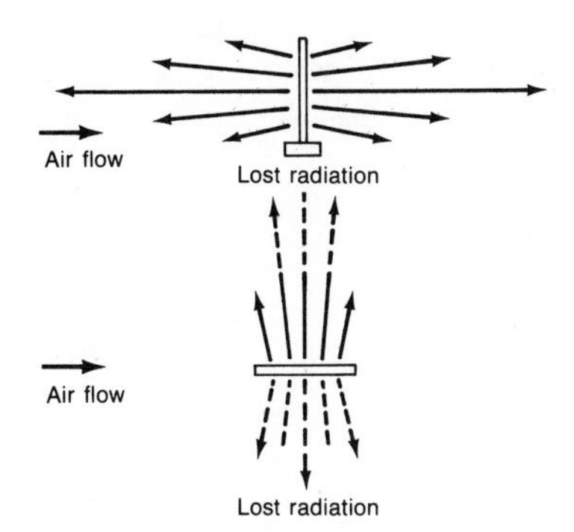

Figure 13–2. Method of installing sterile conditioners in ductwork. (*Courtesy American Ultraviolet Company, Chatham, New Jersey*)

For full irradiation effect (along length of duct) install American Ultraviolet STERILE CONDITIONERS at right angles to air flow. When bank of lamps is used, plane of lamp bank is installed across the air flow.

air. As the principal distribution of the germicidal rays from the lamp is perpendicular to the lamp, installation should be at a point where the rays can travel farthest in the duct or chamber, as shown in Fig. 13–2.

Since the effectiveness of ultraviolet germicidal rays increases with time of exposure, it is also best to install sterile conditioners, when possible, where the air velocity is slowest. The ideal place is in the air chamber after the air filters. Finishing the duct or chamber walls with aluminum paint increases the reflection for maximum utilization of the germicidal ultraviolet rays. Sterile conditioners will form a row across the duct, as shown in Fig. 13–3. Where a large number of lamps is needed, sterile conditioners may be installed in two or more rows, positioned so that lamps are staggered.

Since sterile conditioners require an insertion width of 35 in., it may be

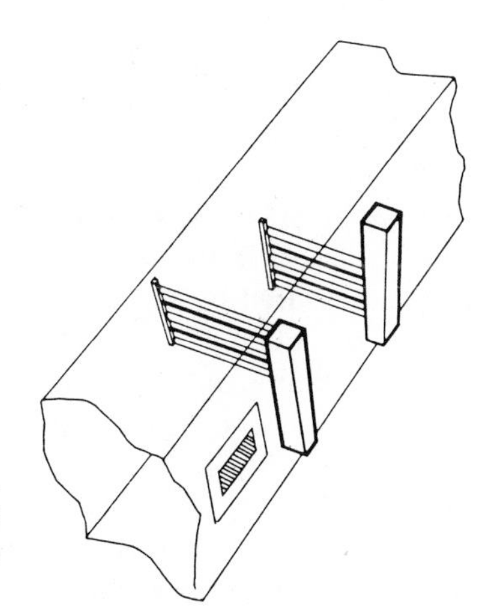

Figure 13–3. Sterile conditioners installed in parallel in duct. (*Courtesy American Ultraviolet Company, Chatham, New Jersey*)

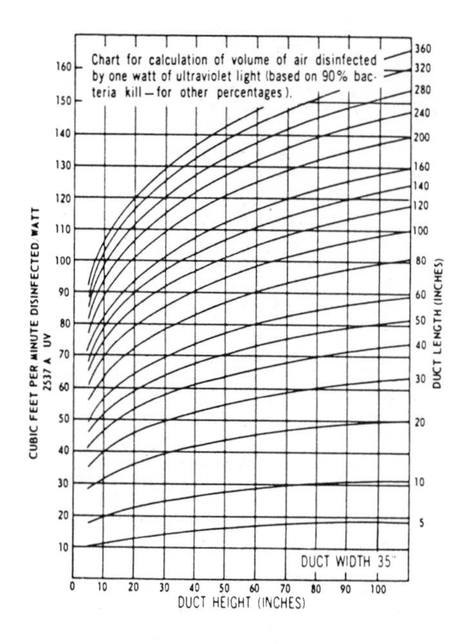

Figure 13–4. Chart for calculation of volume of air disinfected by 1 W of ultraviolet light. (*Courtesy American Ultraviolet Company, Chatham, New Jersey*)

necessary to alter one dimension of the duct to this width. In doing so, the duct should not be made narrow, but large ducts should be 70 in. in width or height so that opposing banks or rows of sterile conditioners may be installed as shown in Fig. 13–4.

MAINTENANCE

The lamps should be cleaned periodically by wiping them with a damp cloth to maintain the ultraviolet output. Frequency of cleaning will vary with the conditions surrounding the installation, but usually once a month is sufficient. Precautions should be taken not to be exposed directly to the ultraviolet rays; otherwise, a "sunburn" of the eyes and skin will result. Therefore, always turn ultraviolet lamps off when servicing.

Lamps incorporated in the system have a rated life of 7500 hours. Lamp efficiency can be measured by use of a special ultraviolet meter. Meters are available from various manufacturers, but adequate ultraviolet output can be assured by replacing lamps on a strict calendar time schedule.

If the sterile conditioner is to be installed in air systems where the ambient duct temperature at the point of installation is below 70°F, special transformers can be supplied in the sterile conditioners to provide proper low-temperature operation of the ultraviolet lamps. It is recommended that the manufacturer be contacted for assistance in such applications.

DESIGN

To determine the number of lamps required and the type of sterile conditioner, first determine the volume of air handled by the duct and the height and length of the duct or irradiation chamber where the sterile conditioners are to be installed. It is assumed that the duct width will be approximately 35 inches, or a multiple thereof, which is the insertion depth or length required for those units manufactured by American Ultraviolet Co. With these dimensions, refer to Fig. 13–4 to find the volume of air disinfected by 1 W of ultraviolet for the particular duct. Find the figure along the right-hand side of the graph which closely corresponds to the length of the duct where the sterile conditioners are to be installed. Follow the curved line adjoining this figure to where it intersects the vertical line that corresponds to the duct height. Directly opposite this point, on the left side of the graph, will be the volume of air disinfected by 1 W of ultraviolet (VDW).

Divide the cubic feet per minute of the duct, at the point of installation, by the VDW to obtain the total watts of ultraviolet required. To find the number of sterile conditioners and types of sterile conditioners required, divide the total watts of ultraviolet required by the ultraviolet output of each lamp as used by the manufacturer.

As lamp output varies with the temperature and speed of the air, the true or corrected ultraviolet output of each lamp of the sterile conditioner should be obtained by referring to Fig. 13–5. Use the curve that closely matches the

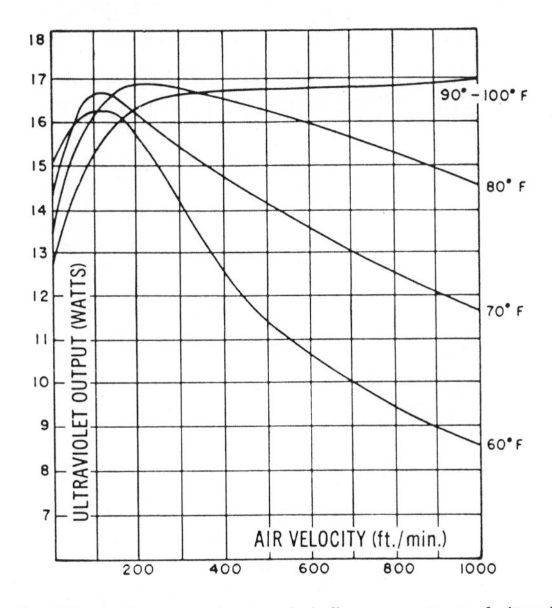

Figure 13–5. Effect of temperature and air flow on output of ultraviolet sterile conditioners. (*Courtesy American Ultraviolet Company, Chatham, New Jersey*)

minimum ambient temperature of the air in the duct. At the point that corresponds to the air velocity (divide the cfm by the cross-sectional area of the duct where installation is to be made), the scale on the left side of the chart will give the ultraviolet output of one sterile conditioner lamp.

The following equation can be used:

$$\text{number of lamps required} = \text{cfm/VDW} \times \text{corrected lamp output}$$

The result is for 90% destruction of bacteria. For higher rates of air disinfection, multiply answer by factor found in Fig. 13-1.

In dealing with an actual application, assume a duct 40 in. high, 100 in. long, and 35 in. wide carrying 10,000 cfm of air at a temperature of 80°F. Referring to Fig. 13-4, we find that 90 ft^3 of air is disinfected by 1 W of ultraviolet (VDW). Applying factors in Fig. 13-5, a sterile conditioner lamp, operating in 80°F, airflow of 1000 ft/min, has an ultraviolet output of 14.6 W. Applying to the formula above,

$$\text{number of lamps} = 10{,}000 \text{ cfm}/90 \times 14.6 = 8 \text{ lamps}$$

The number of lamps required may be determined by splitting the duct into two or three imaginary ducts, each 35 in. wide and applying the same method of calculation as previously described. However, since one large duct with two or three banks of sterile conditioner is more efficient than several one-bank ducts, the result of the calculation must be multiplied by the correction factor found in Fig. 13-6.

Assume a duct 6 ft wide, 5 ft high, and 2 ft long with 30,000 cfm at a temperature of 80°F. Imagine the duct split into two sections, each 3 ft wide, 5 ft high, and 2 ft long, and each drawing 15,000 cfm. Referring to Fig. 13-4, we find that 50 cfm of air will be disinfected by 1 W of VDW.

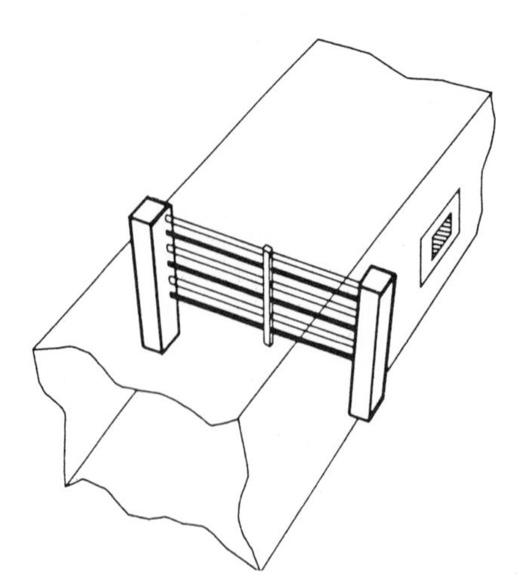

Figure 13-6. Correction factor for two and three banks of lamps. *(Courtesy American Ultraviolet Company, Chatham, New Jersey)*

RESIDENTIAL APPLICATIONS

American Ultraviolet D–10 Series Sterile Conditioners with one to four lamps have been developed for the smaller ducts of residential size and are applicable to practically all residential heating and cooling systems. These units are installed essentially in the same manner as discussed previously, that is, in the center of the longest straight section of duct, perpendicular to the airflow. These units are best employed in the return air chamber. In window or through-wall air conditioners, they will be installed in the exhaust side of the unit. Figure 13–7 shows the possible locations of the D–10 Series Sterile Conditioner in a forced-air heating system.

To calculate the number and type of sterile conditioners required in any given system, the following equation is used:

$$\text{number of lamps required} = \text{cfm/VDW} \times \text{corrected lamp output}$$

The VDW or volume of air disinfected per watt is found in Fig. 13–8. The corrected lamp output (see Fig. 13–9) is dependent on the air velocity passing the sterile conditioner and the temperature of this air. The cubic feet per minute are included in the specifications of each system.

The equation above is based on the 80% kill of microorganisms. Each lamp in the D–10 Series Sterile Conditioners has a controlled production of ozone for minimizing odors.

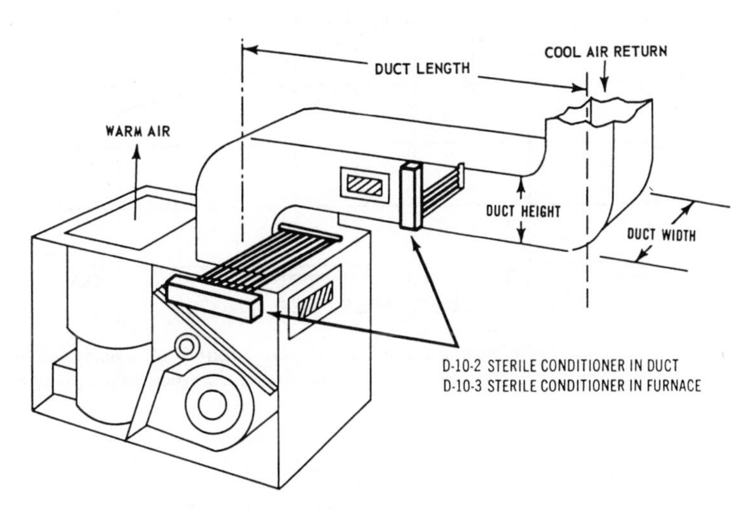

Figure 13–7. Typical residential forced-air heating system. (*Courtesy American Ultraviolet Company, Chatham, New Jersey*)

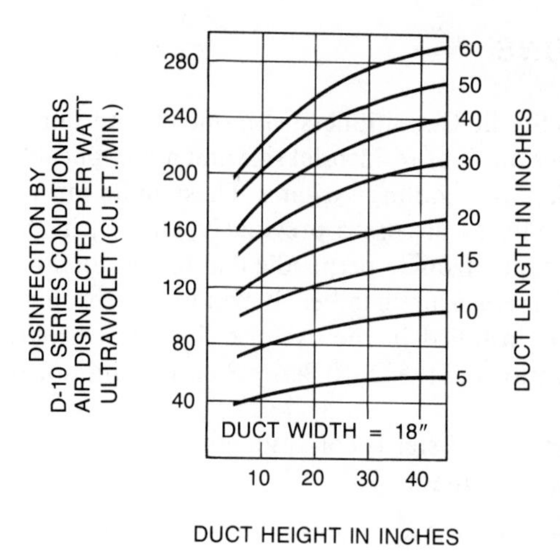

Figure 13–8. Disinfection of D-10 Series per watt of ultraviolet light. (*Courtesy American Ultraviolet Company, Chatham, New Jersey*)

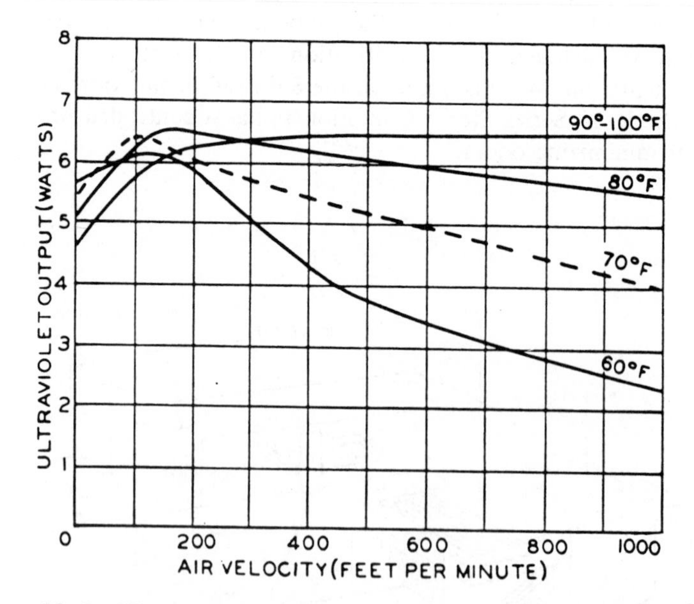

Figure 13–9. Effect of temperature and airflow on output of ultraviolet sterile conditioners. (*Courtesy American Ultraviolet Company, Chatham, New Jersey*)

14

HEAT-SAVING DEVICES

The awareness of accelerated energy costs makes it necessary for almost everyone to save as much energy as possible. The long-range solution to this problem is the use of new technology and energy sources. However, for the immediate future, the solution involves using less energy without sacrificing economic well-being. Various heat-recovery devices have proven to be of benefit to commerical and industrial applications, and some of these may be of some use to homeowners—on a smaller scale, of course. Such devices recover useful energy that would otherwise be lost to the atmosphere.

As much as 50 to 80% of wasted energy can be economically salvaged from heated or cooled exhaust air and then reused. This can be accomplished by a number of methods, which will be discussed in this chapter.

Another important source of heat capture is the collection of energy from outside air, which is then used to heat and cool the inside of a structure. This procedure was covered in Chapter 6.

HEAT WHEELS

There are at least two types of heat wheels in use today. One is the enthalpy exchanger, which recovers both sensible and latent heat; the other is the sensible exchanger, which recovers sensible heat only. Both types require that the exhaust and air-supply ducts be adjacent.

The enthalpy exchanger is installed in the ductwork, as shown in Fig. 14-1. It slowly rotates between the exhaust and supply airstreams. In winter,

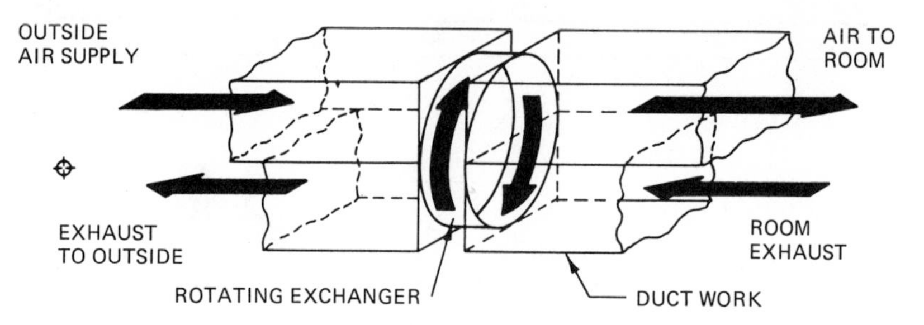

Figure 14–1. Operating principles of an enthalpy exchanger that recovers both sensible and latent heat. (*Courtesy of the author*)

heat and moisture recovered from room exhaust air preheat the incoming outdoor air supply. In summer the reverse occurs as cool and dry conditioned air recovered from the roof exhaust precools the incoming outdoor air supply.

The sensible exchanger operates similarly to the enthalpy exchanger except that it now recovers sensible heat only and is particularly useful for processing applications, high-temperature exhaust, and drying systems.

STATIC HEAT EXCHANGERS

The air-to-air heat exchanger illustrated in Fig. 14–2 represents a static means for transporting heat between the exhaust and outside airstreams which pass through it in a counterflow fashion. It has no moving parts and is maintenance free except in cases where the exhaust air will be cooled to a temperature below dew point; in that case, arrangements for adequate condensate drainage are required.

Like the heat wheel, the static heat exchanger requires that the exhaust and supply air ducts be adjacent to one another (Fig. 14–1) to permit the transfer of heat. The advantages of this unit are

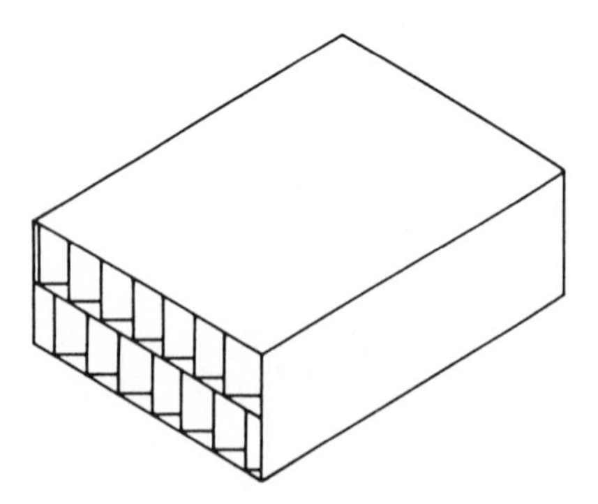

Figure 14–2. Air-to-air heat exchanger. (*Courtesy of the author*)

1. Freedom from maintenance (there are no moving parts)
2. Ruggedness and durability
3. Low installation cost
4. Reduction of heating and cooling loads
5. Efficiency in excess of 70%

HEAT PIPES

The heat pipe is an air-to-air heat-transfer device. It is radically different from most heat exchangers in that it takes advantage of the enormous quantities of heat that are absorbed by a vaporizing liquid or released by a condensing vapor.

A heat pipe consists of a hollow tube which is sealed at both ends. The inside of the tube is a wicking structure consisting of a fine wire mesh and a charge of refrigerant. When one end of the tube is heated, the liquid refrigerant vaporizes, absorbing the heat.

The hot vapor then flows through the hollow center of the wick to the other end of the pipe, releasing the absorbed heat. Figure 14–3 shows the working principles of a heat pipe.

When a stack of heat pipes is installed in a duct system so that the exhaust air flows through one side of the unit and the supply air counterflows through the opposite side, heat is transferred from one airstream to another.

In winter, heat is transferred from the building exhaust air to the cold makeup air, which, in turn, preheats the fresh air. In summer, the process is reversed so that the heat will flow in the opposite direction, that is, from the

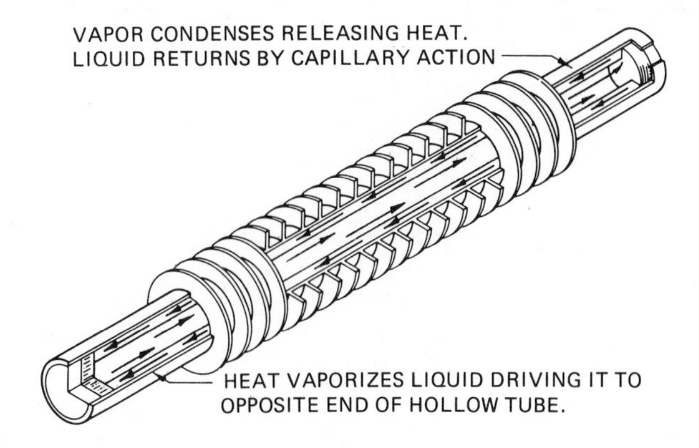

Figure 14–3. Operating principles of a heat pipe (exchanger). (*Courtesy of the author*)

hot makeup air to the cooled building exhaust air, thus precooling the incoming fresh air.

HEAT RECOVERY FROM REFRIGERATION

Frozen-food cases are familiar sights in the modern supermarket. The excess heat given off by the compressors and condensers used in the operation of these cases offers a source of heat which can be captured and returned to the sales area to provide comfort heating and humidity control. This method is probably one of the greatest opportunities for heat recovery and energy savings.

The three basic methods of recovering heat from food service refrigeration are all-air, refrigerant-to-water-to-air, and all-refrigerant. The all-air system is the least complex of the three, and a typical system appears in Fig. 14–4. In this illustration, the return air from the store's sales area is passed through the compressor room, absorbing the rejected heat, which is used to preheat incoming outside ventilation air flowing to the air handler. When the outside air temperature is high enough not to require preheating, the return air is ducted directly to the air handler, bypassing the compressor room.

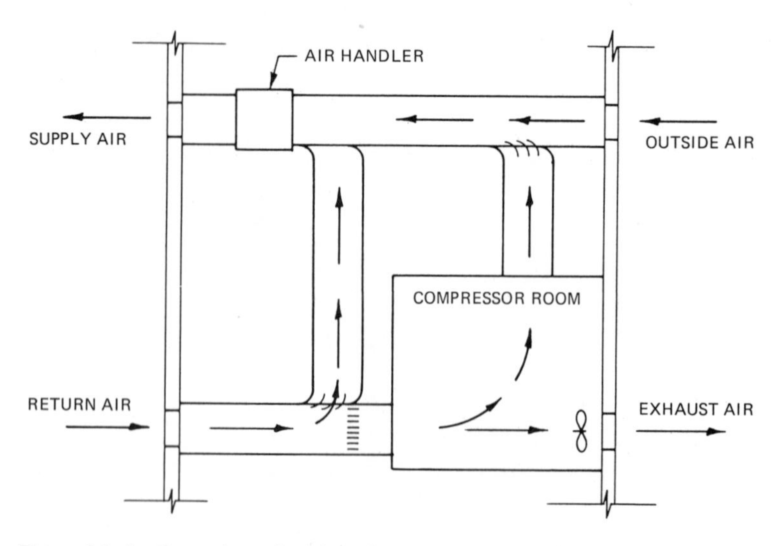

Figure 14–4. Operating principles of an all-air recovery system. (*Courtesy of the author*)

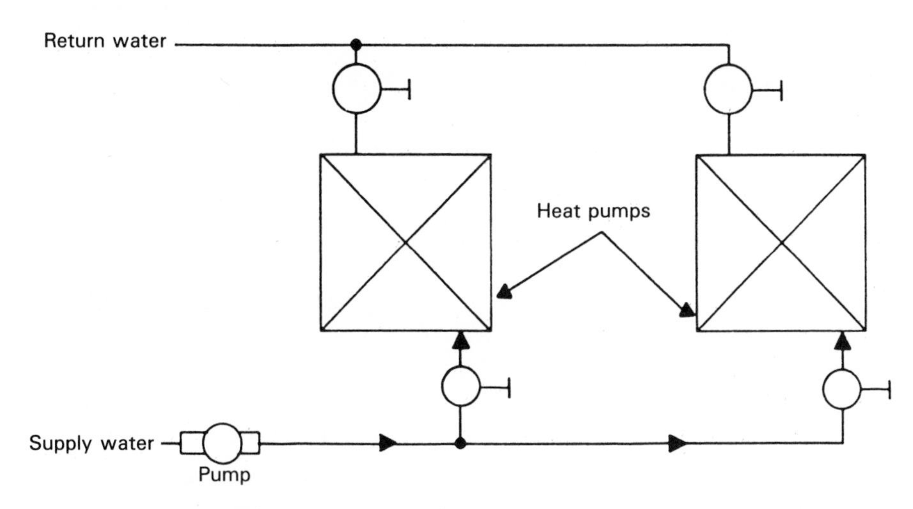

Figure 14–5. Basic operating principles of a water-to-air heat pump. (*Courtesy of the author*)

HEAT PUMPS

The term *heat pump,* as applied to a year-round air-conditioning system, usually denotes a system in which refrigeration equipment is used (1) to extract heat from a heat source and transfer it to a conditioned space when heating is desired, and (2) to remove heat from the space and discharge it to a heat sink when cooling and dehumidification are desired.

Heat pumps operate on several systems; the most common are air-to-air, air-to-water, and water-to-air (Fig. 14.5). All of these systems can provide either heating or cooling, can switch from one to the other automatically as needed, and can supply both simultaneously if desired.

The installation of heat recovery devices offers several advantages. Some of these are:

1. They reduce the operating cost of the heating makeup air in winter.
2. They reduce the operating cost of cooling makeup air in summer.
3. They reduce the size and cost of heating and cooling equipment.
4. They permit the expansion of existing facilities without enlarging existing heating or cooling plants.

15

OIL HEATERS

Many types of oil heaters—other than forced-air furnaces—are used in homes, shops, and other buildings. While the current trend has been turning toward kerosene heaters, regular fuel oil is also used in many models. These heaters also have their share of problems, but in general, proper heating can be had by making sure that the fuel-feeding system is unobstructed, the burner is clean and level, and the fuel is free of water or sediment.

Two types of burners are common in oil stoves: the pot-type burner and the cylinder-type burner. Wick-type burners are common for kerosene heaters, but are not used as often for oil stoves or room heaters. In any of these types, common troubles can be caused by the fuel line being clogged by sediment, or water or air may enter and cause inadequate combustion. Always check the strainer first, as these will usually keep out sediment, but when dirty fuel is used, they will have to be cleaned often. The exact location of these strainers will vary from model to model, but most will be found in the oil basin, under a removable reservoir. In most cases, they can be removed and thoroughly cleaned with AWA 1,1,1 cleaning solution while the stove is in use without disturbing the flow from the lower basin.

The ideal location for strainers is slightly above the bottom of the oil pan, where it will trap small amounts of water, which, being heavier than oil, collect in the strainer and can be dumped out. Another strainer, which is usually located in the lower basin, should also be removed occasionally—but not as often as the top one—and cleaned thoroughly. Many homeowners also like to clean out the basin itself by turning off the oil flow, and while the stove is cold (not lit), dip out as much oil as possible and then sponge out the rest. Again, AWA 1,1,1 cleaning solution can be used for a thorough job.

Fixed-reservoir stoves and those with outside tanks have a constant-level valve to establish the level at which oil enters the burner. Dirt and water can be removed by frequent draining at the sediment trap. The valve contains a cylindrical strainer, which should be rinsed regularly with clean oil after unscrewing the large plug which is located near the bottom in many models. Again, the exact location of this plug may vary slightly from manufacturer to manufacturer. Once this plug has been unscrewed, merely pull it out, clean it, and replace.

Next in line comes air, which can hamper and even stop the flow of oil to the burner. Should a bubble occur, it can sometimes be forced on through the fuel line by a gentle tapping. However, if this does not work, the fuel line will have to be disconnected at the nearest coupling, let the oil flow freely, and then reconnect the line or pipe at the same time, that is, while the fuel is flowing freely.

The vent in the tank cover should be checked by either blowing through it or using a small-diameter wire to clean out all dirt and sediment. However, if this vent is very dirty, it is best to uncouple it at both ends and then flush it out with either clean oil or another approved cleaning solution. Never use gasoline. Many problems will develop from a dent or bend in this tubing, but most can be straightened by running a reaming wire through it and then rebending it very carefully to avoid cracking the tubing or pinching the bore (inside diameter). When replacing this tubing, make sure that there are no high spots where air can be trapped, which, as mentioned previously, will cause trouble.

Leaks are quite common, especially when old tubing is used for the fuel line. Often, these can be cured by a careful tightening of the coupling, but in some cases they cannot. When tightening the coupling does not work, cut off the old joint with a hacksaw, or preferably a tubing cutter, and make a new one. In general, this requires slipping on a new gasket and flaring the tubing slightly with a flaring tool. If no tool is available, usually a piece of tapered metal rod—such as a large drift punch—can be used with good results. Most users of these stoves keep a few extra gaskets on hand, so the minimum of delay will be had when a leak occurs.

An air-locked control valve can usually be freed by opening and closing it several times to let the air bubbles pass through and out of the burner. Some of these valves have an integral cleaning device incorporated in the stem which can be worked up and down or rotated (depending on which model is used) to pass the air out. Water and small particles of sediment may be passed the same way. If the stoppage persists, the valve stem will have to be removed so that it can be cleaned thoroughly. However, before doing so, make absolutely certain that the shutoff valve is closed ahead of the control valve, and that the flame is completely out.

Sediment under the float-mechanism needle in a constant-level valve causes an overflow that trips the automatic safety device and stops the oil flow. Resetting the trip lever will correct a mild case, but if the trouble persists, more experience is required for a correct adjustment.

The burners themselves may be cleaned through the cleanout tube at the point where the fuel line enters the burner. A special tool is usually provided for breaking down the carbon deposit, or a rod may be used. Reasonably good luck can be had with a small stainless steel brush and/or steel wool, but the burner will have to be completely disassembled, which should be done about once a year for the best possible heating from the stove.

Carbon forms in the pot or in the bowl grooves of the burner as a normal result of combustion; but a poorly operating burner may carbon up in a matter of only a few hours, especially if dirty oil is used. When good fuel is used and the burner is operating perfectly, the cleaning of carbon will normally have to be done only once a season. If allowed to go unchecked, carbon can prevent proper distribution and circulation of fuel in the pot or bowl and will also clog the feed pipe, resulting in an irregular flame which is sure to smoke badly. Furthermore, such a flame is hard to regulate and will be highly inefficient. There is also danger of flooding such a burner and losing control of the fire if the burner is forced. Therefore, a burner that shows signs of carbon should be cleaned as quickly as possible.

In many stoves, carbon can be scooped out from the pot (with the flame out of course) by reaching in through the front door with a gloved hand. Others can be cleaned with a large spoon or small scoop after removing the "lighting-hole" plug, that is, the opening where the stove is lit manually. Some of the newer models, however, are equipped with a battery-operated automatic lighter which is activated by either pressing a button—once the fuel valve is turned on—or by pushing a lever. In the manual types, there is usually a gasket of asbestos material sealing this opening. If the plug is removed, make certain that a good gasket is used when the plug is replaced; otherwise, the stove will not fire properly.

Carbon is cleaned from a cylinder or ring-type burner by reaching through the stove door and lifting out the parts in the following order:

1. Lift out the cylinder cover plates.
2. Remove the cylinders.
3. Lift out the vaporizing-chamber cover.
4. Remove the bowl.
5. Remove the kindlers.
6. Remove the vaporizing chamber.

Should the bowl grooves be badly choked, the kindlers must be pried out, and they will probably have to be replaced. Use some type of cleaning tool or the blade of a screwdriver to scrape carbon from the bowl grooves, the ducts running from one groove to another, and the vaporizing chamber. Then run a rod down through the cleanout tube and push out the carbon that has accumulated inside.

Hardened kindlers should be replaced and it is recommended that one extra set always be kept on hand for use when needed. The kindlers should project slightly above the top of the bowl. Cut them to length to fit the circle without overlapping.

Clean out clogged perforations in the cylinders, and replace them, making sure they are properly seated on the bowl by rotating them slightly. If the cylinders are joined with connecting rods, rotate the assembly to ensure seating.

For best results, the burner should be absolutely level so that the oil flow will be equal on all sides and the flame even, and the constant-level valve or reservoir basin to which the burner is connected must also be level. Use a conventional spirit level on the top of the pot from side to side and from front to back. Be sure that the reservoir is full, and make the adjustment by turning the screw-mounted stove feet in or out. If the burner can be reached only through the lighting-plug opening, lay the level on the rails, the rim of the reservoir, or the leveling ledges provided on some stoves. Remove the cylinder assembly and kindlers of a cylinder burner and place the level across the top of the bowl. Some stoves have built-in spirit levels.

If a stove fails to heat properly with the flame turned up as high as it will go, do not experiment with the height setting of the constant-level value unless you are sure that there is no other cause. These valves are factory set at a level that makes flooding the burner practically impossible, and to raise the level may make the stove unsafe to use. It is best to have an experienced service person make this adjustment if at all possible. If not, refer to the owner's manual.

The wick of a wick-type oil and kerosene burner conveys oil to the flame by capillary action, which may be impaired by a hard crust. Frequent brushing with a soft cloth or paper towel will usually keep the wick soft and absorbent. Stoves of this type that are used frequently should be maintained in this matter at least every few days. Remove the flame spreader, raise the wick level to the top of the wick tube, brush off the char, and then pat the wick down evenly. Many manufacturers recommend removing the fuel tank (while the flame is out) or else shut off the shutoff valve so that only the fuel remaining in the pan is available for burning. Take the stove outside, remove the flame spreader, and light the wick. Let the flame burn in this manner until it goes out. This supposedly cleans the wick.

Hard carbon can be scraped from the flange of the spreader with a knife or stainless steel brush, or even steel wool. A stiff brush is almost always necessary when cleaning the carbon from the perforations. Char on the wick tube can be removed with fine abrasive paper. Vents in the burner chimney must also be kept open to provide proper ventilation and to prevent overheating that might crack the procelain-enamel finish on some models. Soap and clean water may be used for most exterior cleaning of the stoves, but should never be cleaned in this manner while still hot.

Any wick that has burned too short should naturally be replaced with a good one. If the new wick cannot be fitted into the burner so that its top is level all around the wick tube, it should be burned off. Drain the reservoir or shut off the fuel supply and remove the oil remaining in the feed pipe by unscrewing its cleanout plug. Then light the burner, adjust the wick to its lowest flame, and allow it to burn dry to level it. After the last spark has died out, resaturate the wick by turning on the oil supply, and then, without changing the position of the wick, remove the chimney and flame spreader and wipe the wick smooth with a cloth or paper towel.

No smoke pipe connection to the house chimney is necessary for a wick stove, but fresh air is necessary if the stove is to burn without smoking. Also, clean fuel is immensely important with this type of burner. When there is any question about the cleanliness of the fuel, there are fuel additives on the market that supposedly remove moisture and impurities. Proper ventilation can be had by leaving a door or window ajar at various intervals.

For sufficient draft in a pot or cylinder burner, have the run of stove pipe as short and direct as possible, and be sure to use a size that fits the flue collar of the stove. Pot burners require a full flue opening at all times. A swinging draft regulator in the piping will help equalize variations in chimney pull. Adjustment of the position of the weight on the regulator blade provides the degree of swinging suitable to the strength of the chimney draft.

Little draft is needed in a cylinder burner after it is once started. A manual damper in the piping will permit the draft to be cut. For both types of stove, the house chimney should be at least 20 feet high; it should be free of cracks and leaks, and all unused openings should be sealed.

When storing an oil stove, be sure to drain out all the kerosene. This keeps water, which is always present to some degree, from settling and causing rust.

A

HEATING AND VENTILATING SPECIFICATIONS*

The specifications for a building or project are the written description of what is required by the owner, architect, or engineer. Together with the drawings, the specifications form the basis of the contract requirements for the construction. The following sample illustrates the general wording and contents of a typical heating and ventilating specification.

DIVISION 15 – MECHANICAL

HEATING AND VENTILATING

1. GENERAL:
   (A) The "Instructions to Bidders," "General Conditions" of the architectural specifications govern work under this Section.
   (B) It is understood and agreed that the Contractor has, by careful examination, satisfied himself as to the nature and location of the work under this Section and all conditions which must be met in order to carry out the work under this Section of the Specifications.
   (C) The Drawings are diagrammatic and indicate generally the locations of materials and equipment. These drawings shall be followed as closely as possible. The architectural, structural, and drawings of other trades shall be checked for dimensions and clearance before installation of any work under this Section. The heating and ventilating contractor shall cooperate with the other trades involved in carrying out the work in order to eliminate interference between the trades.
   (D) The Drawings and Specifications are complementary each to the other, and work required by either shall be included in the Contract as if called for by both.
   (E) It shall be the responsibility of the Heating and Ventilating Contractor to obtain drawings of equipment and materials which are to be furnished by the Owner under separate contract, to which this Contractor is to connect, and for which it is necessary that the Heating and Ventilating Contractor coordinate the work under this Section.
   (F) All services that are interrupted or disconnected shall be rerouted and reconnected in order to provide a complete installation.
   (G) All work shall be performed by mechanics skilled in the particular class or phase of work involved.
   (H) All equipment shall be installed in strict accordance with the respective manufacturer's instructions or recommendations.
   (I) Where appropriate, all equipment shall be UL approved.
   (J) The Heating and Ventilating Contractor shall present five (5) copies of shop drawings or brochures for all fixtures, equipment, and fabricated items to the Engineer for the Engineer's approval. The Heating and Ventilating Contractor shall not proceed with ordering, purchasing, fabricating, or installing any equipment prior to the Engineer's approval of the shop drawings and brochures. Checking is only for general conformance with the design concept of the project and for general compliance with the information given in the contract documents. Any action shown is subject to the requirements of the Plans and Specifications. The contractor is responsible for dimensions that shall be confirmed and correlated at the job site, fabrication processes and techniques of construction, coordination of his work with that of all other trades, and the satisfactory performance of his work.
   (K) All materials and equipment shall be new and undamaged and shall be fully protected throughout the construction period in order that all equipment and materials shall be in perfect condition at the time of acceptance of the building by the Owner. It shall be the responsibility of the Heating and Ventilating Contractor to replace any damaged equipment or materials he is furnishing.

*Courtesy of the author.

(L) The naming of a certain brand or manufacturer in the Specifications is to establish a quality standard for the article desired. This Contractor is not restricted to the use of the specific brand or manufacturer named. However, where a substitution is requested, a substitution will be permitted only with the written approval of the Engineer. The Heating and Ventilating Contractor shall assume all responsibility for additional expenses as required in any way to meet changes from the original materials or equipment specified. If notice of substitution is not furnished to the Engineer within fifteen days after the General Contract is awarded, then equipment named in the Specifications and materials named in the Specifications are to be used.

(M) Excavation and Backfilling:

(1) The Heating and Ventilating Contractor shall do necessary excavation, shoring, and backfilling to complete the work under this Section of the Specifications under the supervision of the General Contractor. No foundation or structural member shall be undermined or weakened by cutting, unless provisions are made to strengthen the member so weakened as necessary.

(2) Excavation shall be cut to provide firm support for all underground conduit, pipe, etc.

(3) Backfill shall be provided as specified under *Backfilling*.

(N) Codes and Standards: All materials and workmanship shall comply with all applicable codes, specifications, industry standards, and utility company regulations.

(1) In cases of differences between the building codes, Specifications, state laws, industry standards, and utility company regulations in the Contract Documents, the most stringent shall govern. The Contractor shall promptly notify the Architect in writing of any such existing difference.

(2) *Noncompliance:* Should the Contractor perform any work that does not comply with the requirements of the applicable building codes, state laws, industry standards, and utility company regulations, he shall bear the cost arising from the correction of the deficiency.

(3) Applicable codes and all standards shall include all state laws, utility company regulations, and applicable requirements of the following nationally accepted codes and standards:

a. National Building Code

b. National Electrical Code

c. Industry codes, Standards, and Specifications:

1. AMCA—Air Moving and Conditioning Association
2. ASHRAE—American Society of Heating, Refrigeration, and Air Conditioning Engineers
3. SMACNA—Sheet Metal and Air Condition Contractors' National Association

(O) This Contractor shall do all cutting and patching of work as necessary to complete the work under this Section. All finished work in conjunction with this work shall be repaired to perfectly match adjoining finished work. This work shall conform to each respective specification section for the particular phase involved. The Heating and Ventilating Contractor shall employ tradesmen skilled in the particular trade involved in carrying out this work. This work shall proceed only with the approval of the Architect. This Contractor shall perform the work under this Section of the Contract in such a way that the amount of cutting and patching of other work shall be kept to an absolute minimum.

(P) This Contractor shall obtain all permits and arrange for all inspections necessary for the installation of this work. All fees in this relation shall be paid for by the Heating and Ventilating Contractor. This Contractor shall provide the Owner with certificates of inspection from all authorities having jurisdiction. This Contractor shall be responsible for notifying the authorities having jurisdiction at each inspection stage, and no work shall progress until the inspection has been completed and the work approved.

(Q) Coordination:

(1) It is called to the Heating and Ventilating Contractor's attention that the ductwork and piping shown on the base bid plan have extremely close clearance.

(2) It shall be the responsibility of the Heating and Ventilating Contractor to coordinate the work under this Section with the Plumbing, and electrical Contractor and other subcontractors.

(3) This Contractor shall check approved equipment drawings of other trades so that all roughing-in work shall be of proper size and in the proper location, for all specific equipment used.

(4) The Heating and Ventilating Contractor shall furnish necessary information to the related trades and shall properly coordinate this information.

(5) The Heating and Ventilating Contractor shall cooperate with the other trades involved in order to eliminate any interference between the trades. This Contractor shall make minor field adjustments to accomplish this.

(6) This Contractor shall cooperate with the other subcontractors in order to establish the responsibilities of each so that work can be completed without delay or interference by this Contractor.

2. WORK INCLUDED:

(A) The work in this Section shall include furnishing all labor, equipment, materials, supplies, and components for complete heating and ventilating systems as indicated on the Drawings and Specifications.

3. ELECTRICAL WORK:

(A) The Heating and Ventilating Contractor is to furnish and attach all necessary components for the heating and control system. The control wiring is to be performed by a qualified electricial contractor employed by the Heating and Ventilating Contractor.

(B) The Heating and Ventilating Contractor is to furnish all motor starters and control components that are specified in the mechanical section of these specifications or shown in schedules on the drawings.

(C) The Electrical Contractor shall furnish all other starters and disconnect switches. The Heating and Ventilating Contractor shall be responsible for the electrical connections and for the electrical wiring between the elements of the pneumatic control system.

(D) The Electrical Contractor shall furnish all conduit, fittings, and materials for the electrical work to connect to the heating and ventilating equipment.

(E) The electrical work required in the Heating and Ventilating Section shall conform to all requirements of the Electrical Section of these Specifications.

(F) All equipment shall be suitable for 208-volt, three-phase, 60-hertz electrical characteristics except motors under ½ hp, which are to be 120-volt, single-phase, 60-hertz. Motors of other characteristics are to be used only where specifically indicated.

4. EQUIPMENT:

(A) Ventilating Units:

(1) Ventilating units shall be the draw-through type as scheduled on the drawings.

(2) Units shall be installed in accordance with the manufacturer's recommendations and be factory assembled and tested.

(3) Casings shall be constructed of steel, reinforced and braced for maximum rigidity. Casings shall be of sectionalized construction and have removable panels for access to all internal parts.

(4) Fans shall be DIDW Class 1 centrifugal and shall be statically and dynamically balanced at the factory. Fan shafts shall run in grease-

lubricated ball bearings, the grease line for internal bearings being brought out to the exterior of the casing.

(5) The interior of the unit casing shall be insulated with 1-inch blanket fiber glass insulation, coated to prevent erosion.

(6) Casings and all accessories, with the exception of the coils described in the following paragraphs, shall be galvanized steel or shall be given a protective baked-enamel finish over a suitable rust inhibitor.

(7) Coils shall have seamless copper tubes and aluminum fins. Fins shall be bonded to tubes by mechanical expansion of the tubes. No soldering or tinning shall be used in the bonding process. Coils shall have a galvanized steel casing no lighter than 16 gauge, shall be mounted so that they are accessible for service, and shall be removable without dismantling the entire unit. Steam coils shall be a nonfreeze, distributing type. Hot-water coils shall be type WS, and the steam coils shall be type NS steam distributing.

(8) Heating coils for hot-water service shall be pitched in the unit casing for proper drainage, and tested at 250 psig air pressure under water.

(9) Each unit shall be provided with an adjustable-speed V-belt drive, having a variable-pitch motor sheath to provide approximately ten-percent variation in speed above and below the factory setting. Drive shall be protected by a suitable guard with openings at fan and motor shafts for use of a tachometer.

(10) Motor shall be general-purpose, squirrel-cage type with same type bearings as air-handling unit. Nameplate rating of motor shall not be exceeded by fan BHP requirements.

(11) See details on drawings for mounting and isolation of units.

(12) Filter boxes shall be furnished with throwaway filters, as scheduled. Filter area shall be such that filter velocity is in accordance with the filter manufacturer's recommendation. Filter boxes shall have access doors on both sides. Filters shall fit snugly to prevent air bypass. Both sides of the filter box shall be flanged for fastener holes. Filter shall be not less than 2 inches thick.

(B) CONVERTOR:

(1) Convertor shall be designed for heating hot water with steam. The unit shall be constructed in accordance with ASME Code and labeled for 150-psi working pressure in shell and tubes.

(2) Materials of construction shall be as follows:

a. Water chamber—cast iron.
b. Shell—carbon steel.
c. Tube sheet—rolled steel.
d. Support and Cradles—steel or cast iron.
e. Tubes—¾ inch O.D. 18-gauge seamless drawn copper.
f. Tube Spaces—soft bronze or copper alloy.

(3) Capacity given in schedule on drawings shall be based on a fouling factor of 0.0005.

(4) A valved ________ No. 60 vacuum breaker shall be provided on shell. At each convertor in water piping, a ________ Series 240 Relief valve, ASME rated and sized for full capacity of convertor, shall be installed and set to relieve at pressure shown on Drawings. Valve shall be located so it cannot be valved off from convertor. Pipe discharge shall be full size to the floor drain.

(5) The Contractor shall provide a stand for support of convertor, constructed of welded structural steel or steel pipe.

(6) Convertor shall be as manufactured by ________ Company, Type SU1.

(C) PUMPS:

(1) Base-mounted pumps shall be ________ "Universal" of vertical split-case design, equipped with mechanical seals for 225°F operating temperature and built for not less than 125-psi working pressure. Motor shall be 1750 RPM, drip proof and specifically designed for quiet operation. The coupler between motor and pump shall be ________ flexible type. Spring-type couplers will not be acceptable. Pump and motor shall have oil-lubricated bronze sleeve bearings. Pump shall have removable bearings frame to permit disassembly of pump without disconnection of piping, hydraulically balanced bronze impeller, and solid cast-iron volute. Pump and motor shall be mounted on a steel base plate.

(2) Base-mounted pumps shall have pump connectors as manufactured by ________ and shall be stainless steel, Type SPCF. They shall be sized the same as the connecting pipe.

(D) CABINET UNIT HEATERS:

(1) Cabinet unit heaters shall be self-contained and factory assembled, and shall consist generally of a filter, heating coil, fan with driving motor, and casing. Heater element shall consist of nonferrous fuses and fins.

(2) Cabinet shall be of steel and be wall mounted or ceiling mounted as indicated on the Drawings.

(3) Front of cabinet shall be removable and not lighter than 16-gauge steel. The back shall be not less than 18-gauge steel. Cabinet shall be provided with steel supports for fan, heating coil, and filter, and shall be steel reinforced members as required to provide a rigid exposed casing and silent operation.

(4) Enclosures shall be complete with inlet and outlet grilles and with doors for access to valves. The element shall be easily accessible for repair after installation of unit. Enclosures shall be galvanized, bonderized, or painted with rust-inhibiting paint at factory, and finished in baked enamel in standard color selected.

(5) Fans shall be centrifugal type with one or more wheels mounted on a single shaft. Fan wheels, shaft, bearings, and fan housing shall be mounted as an integral assembly on a heavy steel mounting plate and securely fastened to enclosures. Bearings of fan and motor shall be self-aligning, permanently lubricated ball type, or sleeve type with ample provision for lubrication and oil reservoirs. Bearings shall be effectively sealed against loss of oil and entrance of dirt.

(6) Filters shall be of the permanent type, mounted in tight-fitting slide-out frames arranged to permit filter renewal or cleaning without removal of front panel.

(7) Heaters shall be tested and rated in accordance with standard test codes adopted jointly by the Unit Heater Division, Air Moving and Conditioning Association, Inc., and the American Society of Heating, Refrigerating, & Air Conditioning Engineers.

(8) Motors shall be resilient mounted on a cushion base, shall be for constant or multispeed operation, and shall have overload protection.

(9) See drawings for capacity, type, and arrangement.

(E) FIN-TUBE RADIATION: Fin-tube heating elements and enclosures, together with required mounting components and accessories, shall be furnished and installed as indicated on plans. Material shall be as manufactured by ______ Radiator Company or approved equal.

(1) *Nonferrous Heating Elements:* Nonferrous heating elements shall consist of full-hard aluminum plate fins, permanently bonded to copper seamless-drawn tube and guaranteed for working pressure at 300°F not less than 150 psi for 1-inch tube and 200 psi for 1¼-inch tube. Fins shall be embedded in copper tube at least .007 inch.

(2) *Enclosures and Accessories:* Enclosures and accessories shall be of style and dimensions indicated on plans and shall be fabricated from electrozinc-coated, rust-resistant, bonderized steel. Enclosures shall be 14 gauge. Enclosure louvers shall be "pencil-proof" type. On wall-to-wall application, enclosures shall be furnished in one piece up to a maximum of 10-feet enclosure length for rooms or spaces measuring a maximum of 10 feet 10 inches in wall length, and a 6-inch end trim shall be used at each end. Enclosures shall be furnished in two or more lengths for wall lengths exceeding 10 feet 10 inches. Corners and end enclosures shall have same method of joining.

End trims, furnished with roll-flanged edges, shall be used between ends of enclosures and walls on wall-to-wall applications. End trims shall be 6-inches maximum length and shall be attached without visible fasteners. Corners and end enclosures shall be furnished where indidicated, shall be the same gauge as enclosures, and shall fit flush with enclosures.

(3) *Enclosure-Supports:* Type-A Dura-Mount Back: Enclosures shall be supported at top by means of a 20-gauge roll-formed continuous mounting channel fabricated from electrozinc-coated, rust-resistant, bonderized steel. There shall be a minimum of eight bends forming strengthening ribs running the length of the channel. The top projection of channel shall position the enclosure ⅞ inch from wall to prevent "gouging" of walls when enclosure is installed or removed. A sponge-rubber dirt sealer shall be provided to the offset section of channel permanently anchored to wall, allowing enclosure installation or removal without disturbing the sealer. Dirt sealers attached to rear flange of enclosure are not permitted.

(4) *Enclosure Brackets and Element Hangers*:

a. Enclosure brackets and element hangers shall be installed not farther than 4 feet. Brackets shall be die-formed from 3⁄16-inch-thick stock, 1½ inches wide, and shall be lanced to support and position lower flange of enclosure. Enclosures shall be firmly attached to brackets by setscrews, operated from under the enclosure. Devices that do not provide positive fastening of enclosures are not acceptable. Brackets shall be inserted in prepunched slots in mounting channel to ensure correct alignment and shall be fastened securely to wall at bottom.

b. Sliding saddles shall support heating element and provide positive positioning of element in enclosure to ensure maximum heating efficiency while preventing any possibility of fin impingement on brackets or enclosure joints during expansion or contraction. Element supports shall be double-saddle design fabricated from 16-gauge electrozinc-coated, rust-resistant, bonderized steel. Saddle shall slide freely on saddle support arm bolted to support bracket. Support arm shall allow 1½-inch height adjustment for pitch. The element support saddle shall allow 1⅝-inch lateral movement for expansion and contraction of heating element. Rod or wire hangers are not acceptable.

(5) *Access Doors:* Access doors shall be provided where noted on plans. Doors shall be at least 8 by 8 inches, shall be located in 12-inch-long access panels, and shall have the same gauge as enclosures.

(6) *Enclosure Dampers:* Enclosure dampers shall be provided where indicated. Damper blades shall be fabricated from 18-gauge electrozinc-coated, rust-resistant, bonderized steel, flanged for added rigidity, and shall be permanently attached to enclosures. Threaded damper screw and trunnion shall provide positive operation of blade in any position between open and closed. Damper shall be operated by knob.

(F) FANS: Fans shall be provided as shown and scheduled on drawings. Care shall be taken when mounting in-line fans between joists to provide access to all parts of the fan. (No fans shall be mounted on roof.)

(1) *In-Line Fans:* All these fans shall be equal to ______ "Square Line Centrifugal Fan" with backward-inclined centrifugal wheel, belt-driven Model SQ. B. See Schedule on drawings.

(2) *Fume Centrifugal Fan:* (Welding Fumes) This fan shall be equal to "ILG" Type PE-Belted with cast-iron wheel. See drawings for capacity and mounting.

(3) *Propeller Fans:* See drawings for location and capacity.

(4) Note: See drawings for mounting and isolation.

(G) UNDERGROUND PIPE CONDUIT: (See Alternate A)

(1) *Scope:* The underground conduit shall include reinforced concrete foundation slabs, unit cast-iron pipe supports, unit sleeve-alignment guides, unit pipe anchors and tile conduit envelope, and shall be insulated as specified in these Specifications.

(2) The tile shall be "Therm-O-Tile," or approved equal, and shall be vitrified and equal to or better than "extra-strength" clay pipe of the same diameter as per ASTM Specification C-278. All joints in the conduit envelope shall be sealed with ______ cement mortar mixed in the proportion of one part ______ cement to two parts clean, sharp sand, and coated after cement has set with "Therm-O-Mastic" compound. Additional weatherproofing consisting of one layer of 30-pound felt shall be applied completely over the tile and lapped over the concrete foundation slab. Laps shall be at least 3 inches and shall be sealed with "Therm-O-Mastic" or hot asphalt.

(3) *Tile Conduit Foundation:* The foundation slab shall be at least 4-inch-thick concrete as specified under Section 3, reinforced with 6 by 6 by No. 10 welded wire mesh, and shall have on emergency drain throughout its run. The foundation slab shall extend through building or manhole walls and shall rest on the masonry

of these walls. The emergency drain shall be continued to building or manhole sumps, to sewers, to dry wells, or to outfalls. Where conduit passes under roadway, it shall be reinforced with an envelope of concrete, 4-inches thick.

(4) *Pipe Supports, Anchors, and Guides:* These accessories shall be as regularly furnished by the manufacturer of the "Therm-O-Tile" conduit and especially designed to fit within the conduit envelope. Location shall be as shown on the Plot Plan. There shall be two guides on each side of each drop.

(5) Conduit shall be sized per the manufacturer's recommendation for the pipe sizes enclosed.

(6) *Shop Drawings:* The successful contractor for the conduit work will be required to furnish, for approval, scale drawings showing cross sections through each separate conduit run and giving all required dimensions of the conduit and centers on which each pipe line will be located within it.

(7) *Wall Openings:* The space around the conduit where it enters walls shall be closed with brick and mortar and completely covered with weatherproof mastic. The space on the inside of the conduit on the inside of building or manhole wall shall be filled with brick and mortar.

(8) *Backfilling:* Backfilling shall be done carefully, and the surface restored to its original condition. Backfilling and tamping shall proceed simultaneously on both sides of the conduit (lengthwise of the trench) until the arch is covered to a depth of at least 18 inches. Rock or stone that might damage the tile shall not be used in the backfill. Backfilling shall comply with Section 2 of the Specifications.

(9) Expansion joints shall be furnished or installed where indicated on Drawings. Model and size shall be as shown on Drawings. Expansion joints shall be as manufactured by ________ Corporation.

(10) Conduit and accessories shall be installed in accordance with the manufacturer's recommendations.

(H) Steam Pressure Reducing Valve Assembly:

(1) *SPRV:* Valve shall be as manufactured by ________ Specialty Manufacturing Corp. Valve body shall be iron and shall have flanged connections. Valve shall be single seated and shall have a stainless-steel valve and seat and deep gland packed stuffing box for minimum friction on valve stem. Valve shall have maximum working pressure of 250 psig, an initial pressure of 75 psig, and a reduced pressure of 10 psig. All parts shall be renewable and interchangeable, and valve seat and disc shall be such that they can be changed without removing valve body from the line. Valve shall be ________ No. 7000 3-inch size, capacity 8600 pounds condensate per hour.

(2) Components of pressure-reducing valves shall be installed as recommended by the manufacturer. See diagram on Drawing M-5.

(3) Safety valve on reduced-pressure side of pressure-reducing station shall be as manufactured by ________ Company, ASME tested and rated, having cast-iron body and 250-pound flanged inlet connection. Valve shall be Catalog No. 630 and shall have capacity indicated on Drawings based on a 3-percent accumulation. Pipe discharge of safety valve shall be full size to exterior of building.

(I) Steam Specialties:

(1) *High-Pressure Steam:* Traps shall be ________ side inlet inverted bucket traps. Traps shall be built for up to 125-psig working steam pressure.

(2) *Low-Pressure Steam:* Traps shall be ________ B series, float and thermostatic type, built for up to 15-psig working steam pressure. Traps shall be sized at ¼-pound of differential.

(3) Traps shall have cast-iron or semisteel body with stainless-steel trim.

(4) Flash tank shall be as detailed on the Drawings.

(J) Heating Water Specialties:

(1) Hot-water expansion tank shall be ________ built for 125-pounds water working pressure, and shall be equipped with proper size "Airtrol" tank fitting and air charger fitting. Tank shall be ________, ASME of size as indicated on drawings.

(2) Air separaters shall be ________ "Rolairtrol." They shall be installed as instructed by manufacturer and shall afford adequate clearance to remove built-in strainer.

(3) Water makeup pressure-reducing valves shall be ________ Type 12, with screwed ends.

(4) Air vents installed on heating coils and at high points in water piping and on all upfeed radiation units shall consist of an air chamber and manual air vent of a type suitable for the particular application where used. See detail on drawings.

(5) *Balancing Fittings:* The Heating and Ventilating Contractor shall furnish and install ________ Series 700 flow meter fittings as a permanent part of the water piping systems for use with ________ No. SD-400-4 flow meter in determining and balancing water flow in all systems. These fittings shall be at points indicated on Drawings and shall be located so as to provide 15 diameters upstream and 5 diameters downstream of uninterrupted straight pipe. The systems shall be balanced as closely as practicable by use of the aforementioned flow meter.

(6) Water relief valves shall be ________ Series 230 or 240 as indicated on Drawings. Valves shall be set to relieve at the pressures indicated on Drawings. Pipe discharge shall be full size to floor drain.

(7) *Thermometers:* Thermometers shall be ________, Type 105 straight form or Type 115-90 degree back angle form, as applicable, industrial type, with standard separable socket, 9-inch scale, 30°F to 240°F range, cast aluminum case, with red-reading mercury tube.

(8) *Pressure Gauges:* Pressure gauges for pipe lines shall be ________ drawn case gauges, No. 1000 for pressure, No. 1002 for vacuum, and No. 1004 for compound. Pressure gauges shall have brass bourdon tube soldered to socket and tip, brass movement, and white-coated metal dial 3½ inches in diameter, graduated to meet system design requirements, normal operating pressure being indicated middial. ________ No. 1106B brass pulsation dampener shall be provided on gauges near pumps and ________ No. 1092 T-handled cock in ¼-inch line to all pressure gauges. No. 1100 pigtail siphon shall be provided on all steam gauges.

(K) Condensate Pump:

(1) One duplex condensation pump shall be furnished and installed where shown on the Plans. The pump shall be manufactured by the ________ Co. The pump shall be driven by open, drip-protected ball-bearing motors, rated for single-phase, 60-hertz, 120-volt ac operation. Each pumping unit shall have capacity at 205°F as indicated. The Heating and Ventilating Contractor shall furnish combination starters with disconnect switch and motor overload protection.

(2) The equipment is to include one cast-iron receiving tank, two pumping units, and accessories listed below:

a. The pumping units shall be bronze fitted throughout, shall be of the centrifugal type, and shall have rotating parts that have been dynamically balanced. The receiving tank, pumping units, and motors shall be assembled by the manufacturer to form an integral unit. A strainer with movable screen is to be installed in the receiving tank.

b. Automatic controls shall consist of a float switch, combination starters, and alternator. Controls shall provide automatic alternation between one pump motor and the other.

(3) If requested, the installing contractor shall secure from the pump manufacturer a factory test report which is to be submitted to the specifying engineer for approval. This report shall show the actual condensate capacity for the pumping units and the power input to the units, all as determined by tests of the actual equipment furnished. The test report is to be certified by the manufacturer as to its correctness and all particulars.

(L) Diffusers, Registers, and Grilles:

(1) *Diffusers:* Diffuser shall be ________ type SFSV with deflectrol, No. 7 finish. Dropped Collar shall be provided where called for on drawings.

(2) *Exhaust Registers:* Exhaust registers shall be ________ type GMRV or GFRV, No. 4 finish. Grille in Dark Room shall be light tight.

(3) *Supply Registers:* Supply registers shall be ________ type GMAV with No. 4 finish.

(4) *Exhaust Grilles:* Exhaust grilles shall be ________ type EMR or GFR with No. 7 finish.

(5) *Door Grilles:* Shall be provided by General Contractor. (See Drawing for location only.)

(M) Mechanical Expansion Joints: Mechanical expansion joints shall be provided as called for on drawings. Size, capacity, and type are given on drawings. Expansion joints and guides shall be equal to ________.

(N) Flexible Connections: Bronze unbraided flexible metal nipples shall be provided to connect to steel heating piping below floor and shall extend from ¾-inch inlet to fin tube radiation above floor. See detail on drawings. Nipples shall be furnished in 18-inch length, ¾-inch hex male end, and suitable for temperatures up to 350°F.

(O) Unit Heaters: Horizontal hot-water–type heaters shall be provided by ________ or equal. See drawing for Model number, size, and capacity.

(P) Fan-Coil Units: Two units of the following type and capacity shall be provided: Model OH 600, 4.2 GPM, 42.2 mbh, 200°F EWT, 180°F LWT, P.D. - 12 feet, 3-speed, and a return-air plenum with permanent filter and a fan control switch. Motor shall have split capacitor, 1/20 hp, 120 volt/1 hp.

(Q) Automatic Temperature Controls:

(1) *General:* The System shall be a complete system of automatic temperature regulation of the pneumatic type with electric accessories and components as indicated. Component parts of the system shall be manufactured by ________; the base bid shall be ________ and can be considered as an alternate at the option of the Contractor. The entire system shall be installed by the Control Manufacturer. All control items except room thermostats shall be properly identified with engraved plastic nameplates permanently attached. Room thermostat locations shall be coordinated to align vertically or horizontally with adjacent light switches or control instruments. Room thermostats shall be 5 feet 6 inches (nominal) above the floor. Room thermostat covers shall be open.

(2) Materials: (Thermostats)

a. Firestats shall be UL approved, manual-reset type T-7602 with an adjustable temperature setting. Range hood shall be set at 250°F; all others at 125°F.

b. Freezestats shall be T-7606 with 20-foot temperature-sensitive element, located downstream from the coil. If any portion of the element senses a temperature below its setting, the contacts shall break.

c. Unit heater thermostats for space mounting shall be T-7162, line-voltage type with SP-ST switching action rated 6 amperes for full load and rated 36 amperes at 120 volts for lock rotor.

d. Surface-mounted aquastats shall be type T-7912 with adjustable set point and 10° differential. Contacts shall be rated 10 amperes at 120 volts.

e. Thermostatic sensors shall be T-5210 style "B" bulbs with 5½-inch and 4-foot capillaries or 8-foot and 17-foot averaging element. Sensors shall be designed to measure a temperature and to convert the measurement to an air-pressure signal that is transmitted to a receiver, a 3- to 15-psi signal. The sensor shall have pneumatic feedback.

f. Fluidic controlling receivers shall be T-9000. The instrument shall accept a 3- to 15-psi pneumatic signal from one or two temperature-relative humidity or pressure transmitters, and have pneumatic feedback and enclosed fluidic circuitry.

g. Receiver controller shall be T-5312 and shall accept a 3- to 15-psi signal from a remote transmitter.

h. Pneumatic thermometer shall be T-5500 to provide visual indication of the temperature. The thermometer shall accept a 3- to 15-psi signal from a remote transmitter. The T-5500 shall be for flush-mounted applications.

i. Day-night thermostats shall be T-4502 equipped with two (2) separate bimetallic elements for day and night operation. The thermostats shall be equipped with an indexing switch for changing the thermostat to day or night operation as desired, or from a control location by changing the air pressure from 15 to 20 pounds.

j. Remote capillary pneumatic thermostats shall be T-8000 for direct or reverse acting. The T-8000 shall be equipped with the proper capillary, *B* bulb for sensing outside air and water temperature and also of the averaging type for all other applications.

k. Remote capillary pneumatic thermostats for two-position applications shall be T-8000.

(3) VALVES:

a. Valves shall be sized by the control manufacturer and shall have threaded connections, except valves over 2 inches shall have flanged connections. Valve packing shall be U-cap silicone except where indicated. Maximum allowable pressure drop shall be 5-feet water column for water valves and 60-percent steam pressure.

b. Valve operators for valve ½ inch to 2 inches shall be V-3000 piston operated. The diaphragm shall be manufactured of Butyl rubber enclosed in a heavy die-cast aluminum housing.

c. Valves for steam or water service shall be V-3752 normally open type. The valve shall be equipped with a V-3000 piston operator. The V-3752 bodies shall be of high-grade cast red brass in sizes ½ inch through 2 inches and shall have a back-setting feature that permits changing the stem packing without interrupting service to the system. The modulating plug shall have a replaceable composition disc, especially compounded for steam or hot water, and shall provide an equal percentage relationship between valve lift and flow at a constant pressure drop.

d. Connector valves shall be V-3800 for steam or water with restrictor mounting space. The valve shall be equipped with a heavy-duty moulded rubber diaphragm of the oval piston type. The valve body shall be of cast red brass in sizes ½ inch through 2 and ¾ inch.

e. Valve for steam or water service shall be V-3970 normally closed type. The valve shall be equipped with a V-3000 piston operator. The V-3870 bodies shall be of high-grade cast red brass in sizes ½ inch through 2 inches and shall have a back-setting feature that permits changing the stem packing without interrupting service to the system. The modulating plug shall have a replaceable compulsion disc especially compounded for steam or hot water, and shall provide an equal percentage relationship between valve lift and flow at a constant pressure drop.

f. Three-way mixing valves shall be V-4322 and ½ inch through 2 inches in size. The body shall be three-way with screwed ends and made of high-grade cast red brass. The body shall be suitable for pressure to 150 psi.

(4) DAMPERS AND DAMPER MOTORS:

a. Automatic control dampers shall be interlocking and airtight. They shall be of opposed-blade construction for modulating service and of parallel-blade construction for two-position service. Dampers shall be of the multilouver construction with brass bearings, channel iron frame, and maximum width of 6 feet.

b. Control dampers shall be D-1200 or D-1300 and manufactured specifically to control the airflow in heating, ventilating, and air-conditioning systems.

c. Frames shall be made of No. 13 galvanized sheet steel, formed into channels and riveted. In addition to the rigid frame construction, corner brackets shall be used to maintain perfect alignment of the damper.

d. Blades shall consist of two formed No. 22 galvanized sheets, spot-welded together for strength to withstand high velocities and static pressures. Square blade pins shall be furnished to ensure nonslip pivoting of the blades when a damper is used as a single module or when it is interconnected with others.

e. Bushings shall be made of oil-impregnated sintered bronze and shall provide constant lubrication.

f. Synthetic elastomer seals shall be provided on the blade edges and on the top, bottom, and sides of the frame, and shall be capable of withstanding air temperatures from —20 to 200°F or from —65 to 400°F, as required for the application. The material and extruded form of the blade edge seals shall create a positive seal when the blades are closed. The seals shall be replaceable if they become damaged. Leakage shall be less than 0.5 percent when closing against 4-inch w.g. static pressure, based on conventional velocity of 2000 FPM.

g. Damper motors shall be provided for all automatic dampers and shall be of sufficient capacity to operate the connected damper. Where required, damper motors shall be equipped with positive positioners.

h. Automatic control dampers are specified to be provided as an integral part of the air units.

(5) PANELS:

a. Control cabinets shall be furnished where specified. In general, it is the intention of this specification that control cabinets be furnished for each air-handling unit, major equipment components, and elsewhere as specified. Control cabinets shall be fabricated of extruded aluminum or steel. The cabinets shall have a face panel for flush-mounting gauges and a subpanel for mounting of controllers, relays, etc. Those controls which require manual positioning or visual indication shall be flush mounted and identified with engraved nameplates on the face panel. The controls that must be accessible for maintenance and calibration only are to be mounted on the subpanel inside cabinet. Each item shall be identified by engraved nameplates.

(6) CONTROL PIPING:

a. Control piping shall be hard-drawn copper tubing where exposed and may be either hard- or soft-drawn copper tubing where it is to be concealed. Either solder or compression fitting shall be used. Tubing shall be run in a neat and workmanlike manner and shall be fastened securely to the building structure. All tubing in finished rooms shall be concealed. Where exposed in unfinished rooms, tubing shall be run either parallel to or at right angles to the building structure. In lieu of copper tubing, plastic tubing may be used in end compartment of unit and in control panel.

b. *Plastic Tubing:* In lieu of copper tubing, high-density virgin polyethylene may be substituted, subject to the following requirements. The tubing shall be rated 600-psi burst pressure at 72°F and 300-psi burst pressure at 140°F. Each tube shall be individually numbered at intervals not exceeding 1 inch. All tubing except tubing in control panels and junction boxes shall be installed in EMT conduit in sizes of ½ inch through 2 inches. Standard electrical fittings shall be used. All bends for conduit 1 inch and larger shall be standard purchased bends. There shall not be more than three 90-degree bends between pull boxes. Termination shall be made in standard electric junction boxes or enclosed control panels. Where connections are made from plastic to copper, protective grommets shall be used to prevent electrolysis. The conduit installation shall conform to the same standard as set forth in the copper piping requirements.

(7) INSTALLATION:

a. EXPOSED:

1. Single polyethylene tubing and soft copper aluminum tubing may be run exposed for a length of 18 inches or less. For lengths that exceed 18 inches, the lines shall be run within enclosed trough or conduit, and this tube carrier system shall be installed in a workmanlike manner, parallel to building lines, adequately supported, etc. All connections, except for terminal connections to valves, damper operators, etc., shall be made inside troughs, junction boxes, or control cabinets.
2. Factory-manufactured bundles of polyethylene tubing, with protective outer sheath and hard copper or aluminum tubing, may be installed without an additional trough or conduit envelope, provided that the tube system is installed in same workmanlike manner as specified above for trough and conduit systems.

b. CONCEALED-ACCESSIBLE:

1. Single polyethylene tubing and soft copper or aluminum tubing, either individual or bundled, shall be installed in a workmanlike manner, securely fastened to fixed members of the building structure at sufficient points to avoid excessive freedom of movement. Field-fabricated bundles shall be tied together with a sufficient number of nylon ties to present a neat, uniform appearance.

c. *Concealed-Inaccessible*: Single polyethylene tubes shall be run within enclosed trough or conduit. Factory-manufactured bundles of polyethylene tubing, with protective outer sheath and soft copper or aluminum tubing, may be installed without an additional trough or conduit envelope. Fitting connections to polyethylene tubing shall not be made within the inaccessible area.

d. *Piping Test:* The piping system shall be tested and made tight under pressure of 30 psi. Leakage will not exceed 10 psi in 12 hours.

(8) Miscellaneous relays, pressure switches, disconnect switches, PE and EP relays, time clocks and other items shall be provided as required for the sequence of control indicated. Time clock shall be seven-day type. The PE relays shall be located within 5 feet of the motor control device.

(9) *Air Compressor:* Air compressor shall be of the electric type complete with tank, gauges, combination pressure-reducing valves, low-pressure relief valve, filter assembly, and necessary accessories. The unit shall be of ample capacity to automatically maintain the desired air pressure with an idle period equal to at least twice that of the operating period. Compressor shall be single-stage, high-pressure (60- to 75-psi) type fitted with galvanized ASME reservoir. Motor shall be provided with built-in overload protection. Compressor and motor shall be sized so that their capacity is sufficient for future addition to building.

(10) WORK BY OTHERS:

a. Dampers and valves will be installed by the Mechanical Contractor.

b. Temperature-control wiring shall be the responsibility of the Temperature Control Contractor. This responsibility shall consist of wiring the following items:

1. P.E. switches
2. E.P. switches
3. Firestats
4. Freezestats
5. Air compressor
6. All interlocking wiring required for the system to function properly.

c. The electrical contractor shall be responsible for all power wiring to the equipment.

(11) SEQUENCE OF OPERATION:

a. *Air-Handling Unit Control:* When the unit fan is started, EP-1 is energized to position the outdoor air damper open through D-1 and through PE-1 on units 7 and 8 which energize their respective exhaust fans. When the unit fan is stopped, PE-1 is de-energized and the outdoor air damper is positioned fully closed. Firestat T-3, located in the filter discharge, stops the fan if the discharge temperature rises above 125°F. T-1, located in the discharge duct, controls V-1 modulater to bypass an increasing amount of water around the coil. T-2, with its element located on the face of the heating coil, prevents the discharge temperature from falling below 35°F by overriding T-1 and modulating V-1. Five (5) units have hot-water coils and require three-way valves; three (3) units are steam coils and require steam valves.

b. *Radiation Control:* During the day operation, T-1 controls V-1 to maintain the desired space temperature. During the night operation, T-1 controls V-1 to maintain a reduced night temperature of 60°F.

c. *Exhaust Fan Control:* T-1, located in the intake of each exhaust fan, stops the unit fan should the intake temperature rise above 125°F.

d. *Pump Bypass Control:* T-1, with its sensor located in the supply and return lines, controls V-1 to maintain the desired differential pump pressure.

e. *Convertor Control:* T-1, with its sensor TT-2 located in the convertor supply, is reset inversely with changes in the outside air temperature by TT-1, to control V-1 and maintain the desired convertor discharge temperature inversely with changes in the outside air temperature. T-2 located in the outside air, stops the pump should the outside air temperature rise above 70°F. When the pump is stopped, EP-1 is de-energized to position V-1, fully closed.

f. *Domestic Hot-Water Control:* T-1, located in the discharge of the convertor, controls V-1 to maintain the desired domestic hot-water temperature. T-2, located in the tank, starts the circulating pump if the temperature rises above a predetermined setting. Aquastat T-3 starts the recirculating pump if the temperature rises above the predetermined setting.

g. *Air Supply:* During the day, seven-day time clock C-1 is energized. It, in turn, energizes EP-1 which positions VA-1 to supply 15 pounds of air to day-night thermostats for day operation. During the night operation, C-1 is de-energized. It, in turn, de-energizes EP-1 which positions VA-1 to supply 20 pounds of air to day-night thermostats for night operation.

h. *Valve and Mechanical Rooms Exhaust-Fan Control:* T-1 cycles the fan to maintain a desired temperature of 85°F. A motorized damper, located at the outside air intake louver, interlocks with fan to open when fan is energized and to close when fan is de-energized. A low-limit thermostat, located downstream of the damper in the outside air duct, shuts off fan if temperature drops below 35°F.

i. *Fan-Coil, Unit-Heater, and Cabinet Unit-Heater Control:* T-1 cycles fans to maintain the space temperature. Aquastat T-2 stops the fans if the supply-water temperature falls below 90°F.

(12) *Service and Guarantee:* The entire control system shall be serviced and maintained in first-class condition by the control manufacturer for a period of one year after acceptance at no extra cost to the Owner. At the end of the one-year guarantee period, the Control Contractor must be capable of furnishing emergency service within a normal requested time.

5. DUCTWORK:

(A) Duct thickness, duct breaking, duct joints (both longitudinal and transverse), duct hangers, and all general ductwork shall be in accordance with the recommendations of "Duct Manual and Sheet Metal Construction for Ventilating and Air-Conditoning Systems," as prepared by the Sheet Metal and Air-Conditioning Contractors' National Association.

(B) Ductwork shall be galvanized steel, manufactured in gauges recommended in the above manual.

(C) Duct fittings shall be equivalent to Air Distribution Institute Standard fittings as a minimum requirement.

(D) Where mains split, splitter dampers must be furnished.

(E) All duct joints are to be airtight at ½-inch water pressure.

(F) *Turning Vanes:* ________ or ________ air-turns are to be used in all elbows except round pipes.

(G) EXTERNAL DUCT INSULATION:

(1) *Exposed Ductwork:* Supply ducts and outside air ducts shall be insulated on the outside with 1-inch-thick J-M No. 814, nonflexible, SPIN-GLAS fiber glass duct insulation. Insulation shall have factory-applied facing—FSK (Foil Skrim Kraft). Insulation shall have an average thermal conductivity not to exceed .23 Btu-inch per square foot per °F per hour at a mean temperature of 75°F. All insulation shall be applied with edges tightly butted. Insulation shall be impaled on pins welded to the duct and secured with speed clips. Pins shall be clipped off close to speed clips. Spacing of pins shall be as required to hold insulation firmly against duct surface, but not less than one pin per square foot. All joints and speed clips shall be sealed with ________ 207 glass fabric set in ________ 30-35, on a 3-inch-wide strip of same facing adhered with ________ 85-20 adhesive.

(2) *Concealed Ductwork:* Supply ducts and outside air ducts shall be insulated on the outside with 1-inch-thick J-M flexible Microlite or approved equal. Insulation shall be cut slightly longer than the perimeter of the duct to ensure full thickness at corners. The insulation shall have an average thermal conductivity not to exceed 0.25 Btu-inch per square foot per °F per hour at a mean temperature of 75°F. All insulation shall be applied with edges tightly butted. Insulation shall be secured with ________ 85-20 adhesive. Adhesive shall be applied so that insulation conforms to duct surface uniformly and firmly. All joints shall be taped and sealed with 3-inch-wide strips of the facing applied with 85-20 adhesive.

(H) Where ducts change shape, enlarge, or reduce, transition is to be made with a maximum angle of 15 degrees, except where it is specifically shown otherwise.

(I) *Volume-Control Duct Dampers and Duct Damper Hardware:* Volume-control duct dampers shall be placed as indicated on drawings and constructed as shown in the SMACNA Duct Manual, Plate No. 28 and multiple-plate volume dampers SMACNA Manual, Plate No. 29. Vent-lok control hardware shall be appropriate for specific use.

(J) FIRE DAMPERS:

(1) Fire dampers shall be installed at locations indicated on the Plans, in full conformance with NFPA Bulletin No. 90-A, and in complete accordance with city, state, and local codes.

(2) All fire dampers shall bear label of UL and be listed under the continuing inspection service of UL, where applicable, and shall have been successfully tested for 1½ hours, up to 1800°F. In mounting conditions not in conformance with UL testing, units shall be built in full conformance with standards of the American Insurance Association and the NFPA Bulletin No. 90-A.

(3) Fire dampers shall be as manufactured by "Fire-Seal" damper. Dampers shall be provided with interlocking blades to form a solid coating of steel when closed. When open, the blades are to be completely concealed in the head of the frame, allowing 100-percent-free undisturbed airflow with minimum turbulence. Entire assembly shall be galvanized for corrosion resistance and shall conform to ASTM Specification A-90-63-T.

(K) *Control Dampers:* Control dampers shall be furnished to the Mechanical Contractor by the Control Manufacturer and shall be installed in accordance with the Control Manufacturer's instructions in the duct system.

(L) *Canvas Connections:* Canvas connections shall be Vent-Fabric "Vent-Fab," 20-ounce waterproof and fireproof canvas approved by Underwriters' Laboratories. Each air-handling unit return and supply shall be connected with canvas connection.

(M) *Sawdust Ductwork:* Sawdust ductwork shall be round, 16 gauge with fittings as shown. Ductwork shall be equal to ________. All joints shall be welded airtight. Welding frame fume exhaust shall be of same material.

(N) Flexible metal exhaust duct shall be equal to ________ or ________.

(O) This Contractor shall provide all louvers equal to Air Balance, 4 inches deep, rain and storm check, and of all-aluminum construction. Installation in wall shall be done by the General Contractor.

(P) *Paint Spray Booth Vent:* Paint spray booth vent shall be fabricated of No. 20 gauge galvanized sheet steel equal to vent manufactured by ________. Vent shall be furnished with a roof flange and weather canopy with rain guard, as manufactured by ________. See drawings for support detail.

6. WIRING DIAGRAMS:

(A) This Contractor shall furnish to the Architect for the Architect's approval five (5) copies of wiring diagrams (i.e., all individual wires diagrammed) showing the complete control wiring system diagram for the heating and ventilating system. The Contractor shall furnish a framed, glass-protected copy of this wiring diagram for the complete system *as installed.* This Contractor shall place this wiring diagram in the Equipment Room.

7. TESTS, CLEANING, AND GUARANTEE:

(A) This Contractor shall provide all pumps, gauges, and other instruments necessary to perform tests as required.

(B) This Contractor shall hydrostatically test the piping of the steam system and heating-water system, to a pressure of 150 percent of the system working pressures. The pressure test shall be for at least 8 hours, at which time pressure shall remain constant without additional pumping. After satisfactorily completing tests and before permanently connecting equipment, this Contractor shall blow and flush piping thoroughly so that interiors of all piping shall be free of foreign matter. All traps, strainers, etc., shall be cleaned at the time of flushing.

(C) The Contractor shall adjust and regulate the completed system under actual heating and ventilating conditions to produce a satisfactory system. All automatic temperature controls shall be adjusted for satisfactory operation during the first heating and ventilating seasons. This Contractor shall make all necessary adjustments during the first heating and ventilating seasons without additional cost to the Owner. (This does not mean that the Heating and Ventilating Contractor is responsible for any negligence of operation by the Owner.)

(A) This Contractor shall furnish complete instructions covering the operation of the heating, ventilating, and control systems.

(B) A framed, glass-protected copy of operating instructions is to be placed in the Mechanical Equipment Room by this Contractor.

9. PERFORMANCE TESTS:

(A) This Contractor shall provide all necessary instruments to perform tests as required.

(B) The Heating and Ventilating Contractor shall conduct the following tests upon completion of installation of the system under the direction of the Architect.

(1) *Air Distribution System:* Performance test after proper balancing of system, showing airflow measurements through each supply and return.

(2) *Ventilation System:* Performance test after proper balancing, showing airflow measurements through each exhaust grille.

(3) *Heating System:* Operating test of the entire system during cold weather with fiinal adjustment to outdoor design conditions as necessary and to the heating system. In the event the weather prevents testing before acceptance of the building, the building will be accepted subject to successful completion of the above tests.

(4) *Air-Handling Equipment:* Airflow tabulation and listing of outlet and inlet temperatures and heating medium inlet-outlet temperatures.

(C) Performance tests shall include the complete heating and ventilating systems and all their parts, including thermostatic and electrical controls, in order to determine that the systems are in compliance with the Contract. Tests shall show that the heating and ventilating systems are acceptable before the installation is approved for acceptance by the Owner. The Contractor shall furnish the Owner with four (4) copies of the finds of the approved texts, including tabulation of all readings for the job and computations. Final payment to the Contractor, less an amount to cover the cost of certain tests, shall be made when weather conditions have caused postponement of certain tests.

(D) The heating units shall be checked for performance to determine that the Specifications are met in every respect.

(E) This Contractor shall guarantee all materials and equipment installed by him, and all workmanship for a period of one year after final acceptance of the heating work against all defects occurring during that period.

(F) The heating and ventilating system shall be tested under operating conditions for a period of five 8-hour days or as necessary to demonstrate that the requirements of the Contract are fulfilled.

(G) This Contractor shall adjust and regulate the completed system under actual heating conditions to produce a satisfactory system. All automatic temperature controls shall be adjusted for satisfactory operation in the first heating season. This Contractor shall make all necessary adjustments in the first heating season without additional cost to Owner. (This does not mean that the Heating Contractor is responsible for any negligence in operation by Owner.)

10. MARKING OF CONTROLS, VALVES, AND ELECTRIC CONTROLS:

(A) All valves, electric controls, and electric starters and switches are to be marked with permanent metal tags. The tags shall be black enamel finish with engraved, white enameled letters that are at least ¼ inch high.

(B) The location at ceiling of all damper electric controls and balancing valves and duct balancing dampers shall be marked above the ceiling with tag, as described above and screwed to the metal ceiling runners at each location.

11. TEMPORARY HEATING:

(A) The General Contractor shall be responsible for furnishing temporary heating throughout the period of construction of the building.

12. PAINTING:

(A) Factory-painted equipment and materials shall receive primer coat and two coats of enamel, factory applied.

(B) Factory-painted equipment shall be touched up as necessary and where factory-painted color does not match the Architect's color scheme in finished areas.
(C) Factory-primed equipment and materials shall be painted with two coats of enamel in accordance with the Painting Section of these Specifications. This paint shall be applied by the Painting Contractor who shall use experienced painters for this purpose in accordance with the Painting Section of these Specifications.
(D) Unpainted equipment and materials shall receive one coat of primer and two coats of enamel in accordance with the requirements of the Painting Section of these Specifications for exposed metal. This paint shall be applied by the Painting Contractor who shall use experienced painters for this purpose in accordance with the Painting Section of these Specifications.
(E) The inside of all ductwork visible in finished building shall be painted dull black, the paint to be applied in accordance with Painting Section of the Specifications.
(F) All colors shall be as approved by the Architect.

13. TEMPORARY POWER:
(A) The General Contractor shall provide, as required, 208/120-volt, single-phase, 60-hertz temporary power for use of the Heating and Ventilating Contractor.

14. LUBRICATION PRIOR TO START-UP:
(A) Prior to start-up, the Heating and Ventilating Contractor shall fully lubricate all equipment under this Section in accordance with the respective manufacturers' recommendations.

B

SUPPLY AND RETURN AIR OUTLETS

Manufacturers of diffusers, registers, and grilles usually provide tables that show the proper size according to CFM requirements. Typical examples are shown herein. All Tables courtesy Borg-Warner Air Conditioning, Inc.

Table B-1. Side Wall Registers

CFM	Size	Throw-Feet
60	10 x 4	10
80	10 x 6	12
100	14 x 6	14
120	14 x 6	16
140	16 x 6	16
160	16 x 6	18
180	20 x 6	14
200	20 x 6	16
220	20 x 6	18
240	20 x 6	20
260	24 x 6	20
280	30 x 6	18
300	30 x 6	20

Table B-2. Round Ceiling Diffusers

CFM	Diameter (in)
100	6
120	8
140	8
160	8
180	8
200	10
220	10
240	10

Table B-3. Floor Outlets

CFM	Size	Spread (ft)
60	2¼ x 10	7
80	2¼ x 12	11
100	2¼ x 14	11
120	4 x 10	11
140	4 x 10	13
160	4 x 12	13

Table B-4. Return Air Grilles

CFM	Free Area Sq in	Side Wall Return Grilles	Floor Grilles
60- 140	40	10 x 6	4 x 14
140- 170	48	12 x 6	4 x 18 or 6 x 10
170- 190	55	10 x 8	4 x 18 or 6 x 12
190- 235	67	12 x 8	6 x 14
235- 260	74	18 x 6	6 x 16 or 8 x 14
260- 370	106	12 x 12	8 x 20
370- 560	162	18 x 12	8 x 30
560- 760	218	24 x 12	10 x 30 or 12 x 24
760- 870	252	18 x 18	12 x 30
870- 960	276	30 x 12	12 x 30
960-1170	340	24 x 18	
1170-1470	423	30 x 18	18 x 30 14 x 30
1470-1580	455	24 x 24	20 x 30
1580-1770	510	36 x 18	22 x 30
1770-1990	572	30 x 24	24 x 30
1990-2400	690	36 x 24	24 x 36
2400-3020	870	36 x 30	30 x 36

Table B-5. Recommended and Maximum Velocities For Residences Duct

	Recommended Velocities (FPM)	Maximum Velocities (FPM)
Filters (total face area)	250	300
Main Ducts (net free area)	700-900	800-1200
Branch Ducts (net free area)	600	700-1000
Branch Risers (net free area)	500	650-800

Table B-6. Recommended Return Intake Face Velocities

Intake Location	Velocity Over Gross Area (FPM)
Above occupied zone	800-Up
Within occupied zone, not near seats	600-800
Within occupied zone, near seats	400-600
Door or wall louvers	200-300
Undercutting of doors (through undercut area)	200-300

Table B-7. Rectangular Main Ducts 6-in Size

		Recommended (800 FPM)		Maximum (1000 FPM)	
6-In Size	Equiv Dia in Inches	Approx CFM	Approx Inches-Static Pressure Drop Per 100 Equivalent Feet	Approx CFM	Approx Inches-Static Pressure Drop Per 100 Equivalent Feet
4 x 6	5.3	120	.25	150	.35
5 x 6	6.0	160	.20	200	.30
6 x 6	6.8	200	.17	250	.26
8 x 6	7.5	250	.15	310	.23
10 x 6	8.4	310	.13	380	.20
12 x 6	9.1	360	.12	450	.18
14 x 6	9.8	420	.11	520	.16
16 x 6	10.4	460	.105	600	.15
18 x 6	11.0	530	.094	650	.14
20 x 6	11.5	590	.089	710	.135
22 x 6	12.0	630	.084	790	.13
24 x 6	12.4	680	.080	840	.125
26 x 6	12.8	710	.076	900	.120
28 x 6	13.2	750	.073	940	.115
30 x 6	13.6	800	.071	1000	.110
32 x 6	14.0	850	.069	1090	.105
34 x 6	14.4	900	.067	1170	.100
36 x 6	14.7	940	.065	1200	.096

Table B-8. Rectangular Main Ducts 8-in Size

		Recommended (800 FPM)		Maximum (1000 FPM)	
8-In Size	Equiv Dia in Inches	Approx CFM	Approx Inches-Static Pressure Drop Per 100 Equivalent Feet	Approx CFM	Approx Inches-Static Pressure Drop Per 100 Equivalent Feet
4 x 8	6.1	160	.19	200	.29
5 x 8	6.9	210	.165	260	.25
6 x 8	7.5	250	.15	300	.23
8 x 8	8.8	340	.13	420	.19
10 x 8	9.8	420	.11	520	.17
12 x 8	10.7	500	.095	630	.145
14 x 8	11.5	580	.089	710	.135
16 x 8	12.2	650	.083	810	.125
18 x 8	12.9	715	.078	900	.118
20 x 8	13.5	790	.073	990	.110
22 x 8	14.1	870	.069	1100	.105
24 x 8	14.6	930	.066	1190	.100
26 x 8	15.2	1000	.063	1290	.095
28 x 8	15.6	1080	.060	1375	.090
30 x 8	16.1	1150	.058	1450	.086
32 x 8	16.5	1200	.056	1500	.085
34 x 8	17.0	1250	.055	1590	.083
36 x 8	17.4	1320	.054	1620	.080

Table B-9. Rectangular Main Ducts 10-in Size

10-In Size	Equiv Dia in Inches	Recommended (800 FPM)		Maximum (1000 FPM)	
		Approx CFM	Approx Inches-Static Pressure Drop Per 100 Equivalent Feet	Approx CFM	Approx Inches-Static Pressure Drop Per 100 Equivalent Feet
4 x 10	6.8	200	.170	260	.260
6 x 10	8.4	310	.130	380	.200
8 x 10	9.8	420	.110	520	.170
10 x 10	11.0	520	.093	670	.140
12 x 10	12.1	630	.081	800	.130
14 x 10	13.0	730	.075	920	.120
16 x 10	13.8	820	.071	1020	.110
18 x 10	14.8	940	.064	1200	.098
20 x 20	15.5	1050	.061	1325	.092
22 x 10	16.0	1150	.058	1400	.088
24 x 10	16.7	1250	.055	1500	.084
26 x 10	17.3	1350	.053	1650	.080
28 x 10	17.8	1425	.052	1725	.078
30 x 10	18.4	1500	.049	1800	.074
32 x 10	19.0	1575	.048	1950	.071
34 x 10	19.5	1650	.046	2050	.070
36 x 10	20.0	1725	.044	2150	.068
40 x 10	21.0	1900	.042	2300	.064
44 x 10	21.8	2100	.040	2600	.061
48 x 10	22.8	2250	.038	2900	.058
54 x 10	23.8	2400	.036	3100	.055
60 x 10	25.0	2800	.034	3400	.051

Table B-10. Rectangular Main Ducts 12-in Size

12-In Size	Equiv Dia in Inches	Recommended (800 FPM)		Maximum (1000 FPM)	
		Approx CFM	Approx Inches-Static Pressure Drop Per 100 Equivalent Feet	Approx CFM	Approx Inches-Static Pressure Drop Per 100 Equivalent Feet
4 x 12	7.4	240	.150	300	.230
6 x 12	9.1	370	.120	450	.180
8 x 12	10.7	500	.095	630	.145
10 x 12	12.1	630	.081	800	.130
12 x 12	13.0	720	.075	930	.120
14 x 12	14.3	880	.068	1100	.100
16 x 12	15.2	1000	.062	1250	.095
18 x 12	16.1	1150	.058	1450	.088
20 x 12	17.0	1250	.055	1600	.082
22 x 12	17.8	1350	.052	1750	.078
24 x 12	18.4	1450	.049	1850	.073
26 x 12	19.2	1600	.048	2000	.070
28 x 12	19.8	1700	.046	2150	.068
30 x 12	20.3	1800	.044	2250	.066
32 x 12	21.0	1900	.042	2400	.064
34 x 12	21.5	2000	.041	2500	.061
36 x 12	22.2	2100	.039	2700	.058
40 x 12	23.2	2300	.037	2900	.056
44 x 12	24.0	2500	.036	3150	.054
48 x 12	25.3	2800	.033	3500	.050
54 x 12	26.5	3100	.032	3850	.047
60 x 12	27.8	3400	.030	4150	.045

Table B-11. Round Branch Ducts

Dia Size in Inches	Recommended (600 FPM)		Maximum (900 FPM)	
	Approx CFM	Approx Inches-Static Pressure Drop Per 100 Equivalent Feet	Approx CFM	Approx Inches-Static Pressure Drop Per 100 Equivalent Feet
4	52	.20	80	.40
5	82	.15	125	.31
6	118	.12	180	.25
7	160	.095	240	.20
8	210	.080	320	.17
9	270	.070	400	.15
10	323	.060	490	.15
12	470	.050	700	.105

Table B-12. Branch Risers

Size	Equiv Dia in Inches	Recommended (500 FPM)		Maximum (700 FPM)	
		Approx CFM	Approx Inches-Static Pressure Drop Per 100 Equivalent Feet	Approx CFM	Approx Inches-Static Pressure Drop Per 100 Equivalent Feet
2¼ x 12	5.3	76	.098	108	.18
3¼ x 8	5.5	83	.090	118	.17
3¼ x 10	6.0	98	.082	138	.15
3¼ x 12	6.5	114	.077	158	.14
3¼ x 14	7.2	140	.065	200	.12

Notes
1. Do not interpolate static pressure drop between different sizes of ducts.
2. It is permissable to interpolate between recommended and maximum columns for static pressure drop for the same size duct if CFM falls between the CFM's shown.

Table B-13. Rectangular Return Ducts 6-in Size

6-In Size	Equiv Dia in Inches	Recommended (500 FPM)		Maximum (700 FPM)	
		Approx CFM	Approx Inches-Static Pressure Drop Per 100 Equivalent Feet	Approx CFM	Approx Inches-Static Pressure Drop Per 100 Equivalent Feet
4 x 6	5.3	75	.095	110	.180
5 x 6	6.0	95	.081	140	.150
6 x 6	6.8	130	.070	180	.130
8 x 6	7.5	150	.062	210	.120
10 x 6	8.4	190	.053	260	.100
12 x 6	9.1	220	.048	320	.090
14 x 6	9.8	275	.043	375	.083
16 x 6	10.4	310	.040	410	.080
18 x 6	11.0	330	.038	450	.073
20 x 6	11.5	370	.036	510	.068
22 x 6	12.0	390	.035	550	.065
24 x 6	12.4	410	.033	590	.062
26 x 6	12.8	450	.032	620	.060
28 x 6	13.2	475	.031	650	.058
30 x 6	13.6	500	.030	700	.055
32 x 6	14.0	520	.029	750	.053
34 x 6	14.4	550	.028	780	.051
36 x 6	14.7	580	.027	800	.050

Table B-14. Rectangular Return Ducts 8-in Size

8-In Size	Equiv Dia in Inches	Recommended (500 FPM) Approx CFM	Recommended (500 FPM) Approx Inches-Static Pressure Drop Per 100 Equivalent Feet	Maximum (700 FPM) Approx CFM	Maximum (700 FPM) Approx Inches-Static Pressure Drop Per 100 Equivalent Feet
4 x 8	6.1	100	.080	140	.150
5 x 8	6.9	130	.070	180	.130
6 x 8	7.5	150	.062	210	.120
8 x 8	8.8	210	.050	290	.095
10 x 8	9.8	275	.043	360	.084
12 x 8	10.7	320	.040	440	.074
14 x 8	11.5	360	.037	500	.070
16 x 8	12.2	400	.035	570	.064
18 x 8	12.9	450	.032	630	.060
20 x 8	13.5	490	.031	690	.056
22 x 8	14.1	550	.029	760	.052
24 x 8	14.6	575	.027	810	.051
26 x 8	15.2	610	.027	870	.049
28 x 8	15.6	680	.025	910	.048
30 x 8	16.1	710	.024	1000	.045
32 x 8	16.5	750	.023	1050	.043
34 x 8	17.0	780	.022	1100	.042
36 x 8	17.4	810	.022	1150	.041

Table B-15. Rectangular Return Ducts 10-in Size

10-In Size	Equiv Dia in Inches	Recommended (500 FPM) Approx CFM	Recommended (500 FPM) Approx Inches-Static Pressure Drop Per 100 Equivalent Feet	Maximum (700 FPM) Approx CFM	Maximum (700 FPM) Approx Inches-Static Pressure Drop Per 100 Equivalent Feet
4 x 10	6.8	135	.070	175	.130
6 x 10	8.4	190	.053	260	.100
8 x 10	9.8	275	.043	360	.084
10 x 10	11.0	330	.038	460	.072
12 x 10	12.1	400	.034	570	.063
14 x 10	13.0	470	.032	640	.058
16 x 10	13.8	510	.029	720	.053
18 x 10	14.8	600	.027	830	.050
20 x 10	15.5	650	.025	920	.047
22 x 10	16.0	700	.024	980	.045
24 x 10	16.7	750	.023	1050	.043
26 x 10	17.3	810	.022	1150	.041
28 x 10	17.8	880	.021	1200	.039
30 x 10	18.4	920	.021	1300	.038
32 x 10	19.0	975	.020	1400	.037
34 x 10	19.5	1025	.019	1450	.036
36 x 10	20.0	1100	.018	1500	.035
40 x 10	21.0	1200	.017	1700	.033
44 x 10	21.8	1300	.017	1800	.032
48 x 10	22.8	1400	.016	1950	.030
54 x 10	23.8	1500	.015	2150	.028
60 x 10	25.0	1750	.014	2300	.026

Table B-16. Rectangular Return Ducts 12-in Size

12-In Size	Equiv Dia in Inches	Recommended (500 FPM) Approx CFM	Recommended (500 FPM) Approx Inches-Static Pressure Drop Per 100 Equivalent Feet	Maximum (700 FPM) Approx CFM	Maximum (700 FPM) Approx Inches-Static Pressure Drop Per 100 Equivalent Feet
4 x 12	7.4	145	.064	210	.120
6 x 12	9.1	220	.048	320	.090
8 x 12	10.7	320	.040	440	.074
10 x 12	12.1	400	.034	570	.063
12 x 12	13.0	460	.031	630	.058
14 x 12	14.3	530	.028	780	.051
16 x 12	15.2	620	.026	880	.048
18 x 12	16.1	710	.024	1000	.044
20 x 12	17.0	790	.023	1100	.041
22 x 12	17.8	880	.022	1200	.040
24 x 12	18.4	910	.021	1300	.038
26 x 12	19.2	990	.020	1400	.036
28 x 12	19.8	1050	.019	1450	.035
30 x 12	20.3	1125	.018	1550	.034
32 x 12	21.3	1200	.018	1650	.033
34 x 12	21.5	1275	.017	1750	.032
36 x 12	22.2	1325	.017	1850	.031
40 x 12	23.2	1450	.016	2050	.029
44 x 12	24.0	1500	.015	2150	.028
48 x 12	25.3	1700	.014	2450	.026
54 x 12	26.5	1900	.013	2650	.024
60 x 12	27.8	2100	.012	2900	.023

Table B-17. Round Return Branch Ducts

Diameter in Inches	Recommended (500 FPM) Approx CFM	Recommended (500 FPM) Approx Inches-Static Pressure Drop Per 100 Equivalent Feet	Maximum (700 RPM) Approx CFM	Maximum (700 RPM) Approx Inches-Static Pressure Drop Per 100 Equivalent Feet
4	42	.140	60	.250
5	68	.105	95	.190
6	98	.081	140	.160
7	135	.069	185	.130
8	175	.058	250	.110
9	220	.050	310	.092
10	275	.045	380	.080
12	390	.035	550	.065

Table B-18. Return Air Grille and Register Sizing

Size or Equivalent	Free Area (Sq Ft)	Air Capacities CFM Recommended Velocity (500 FPM)	Air Capacities CFM Maximum Velocity (700 FPM)
10 x 6	.275	137	192
12 x 6	.336	168	235
10 x 8	.382	191	267
12 x 8	.467	233	327
18 x 6	.517	258	362
12 x 12	.730	365	511
18 x 12	1.120	560	784
24 x 12	1.517	758	1062
18 x 18	1.743	871	1220
30 x 12	1.911	955	1338
24 x 18	2.336	1168	1635
30 x 18	2.942	1471	2059
24 x 24	3.156	1578	2209
36 x 18	3.549	1774	2484
30 x 24	3.974	1987	2782
36 x 24	4.792	2396	3354

C

DUCT SIZING CHART

Example of Use:

A small utility room requires 35 CFM of air at a velocity of 400 fpm. What size duct must be used to deliver this volume of air?

1. Find 35 CFM on the CFM scale at the left of the chart.
2. Move horizontally to the right, to the 400 fpm line.
3. At the intersection of 35 CFM and 400 fpm, locate the duct diameter line.
4. Read 4-inch duct diameter.
5. Move downward from the intersection to the friction loss scale and read .09-inch water friction loss.

Chart C-2. Equivalent Rectangular Duct Chart (*Courtesy of the author*)

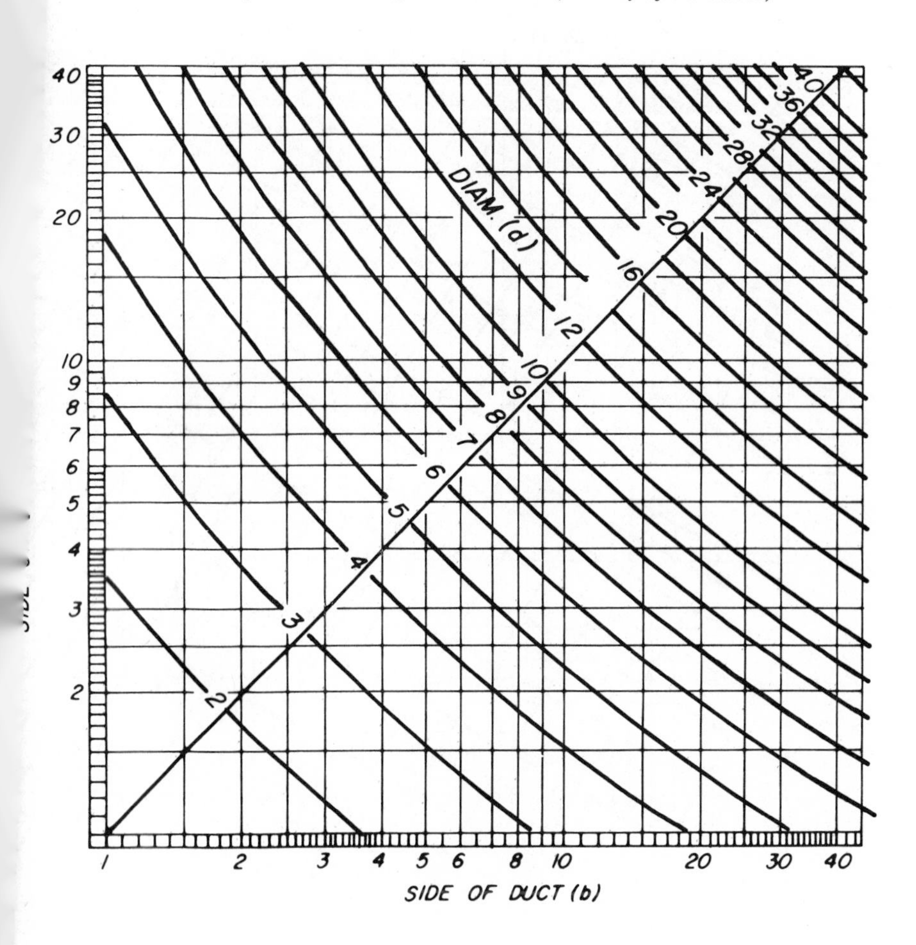

Chart C-1. Duct Sizing Chart (*Courtesy of the author*

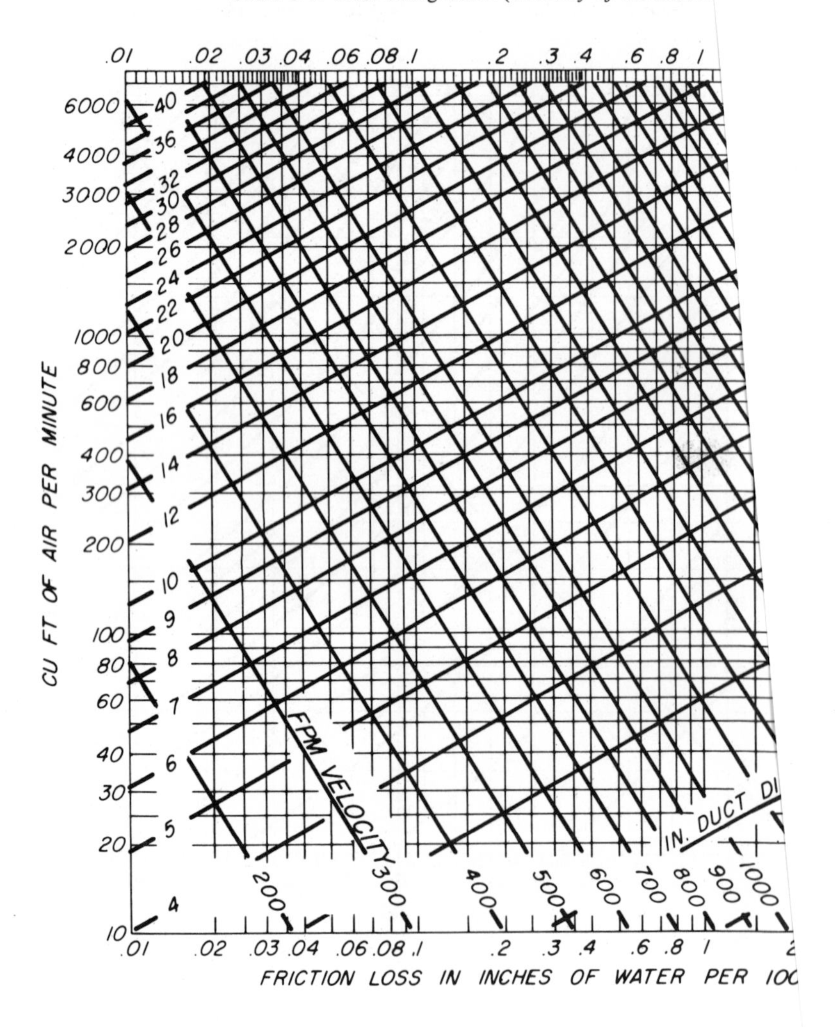

Chart C-3. Use of Duct Sizing Chart (*Courtesy of the author*)

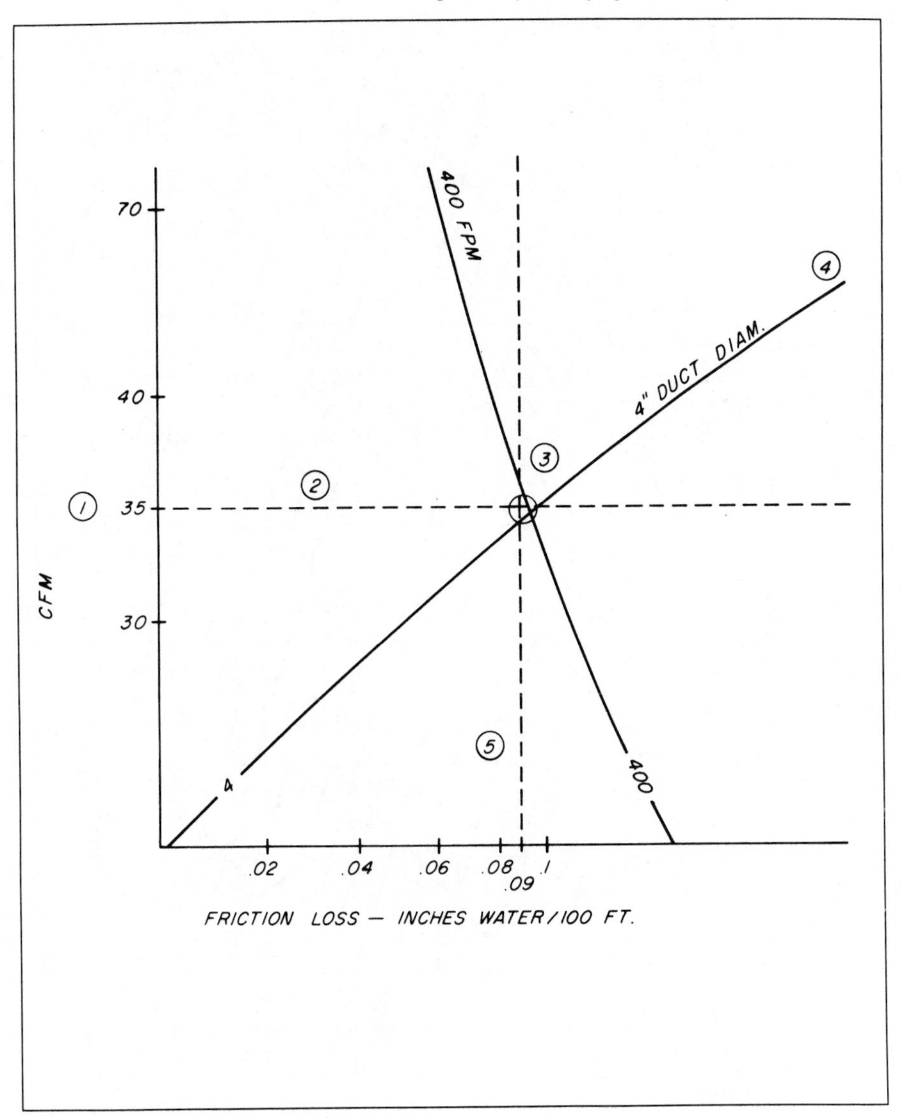

Chart C-4. Friction Loss Chart (*Courtesy of the author*)

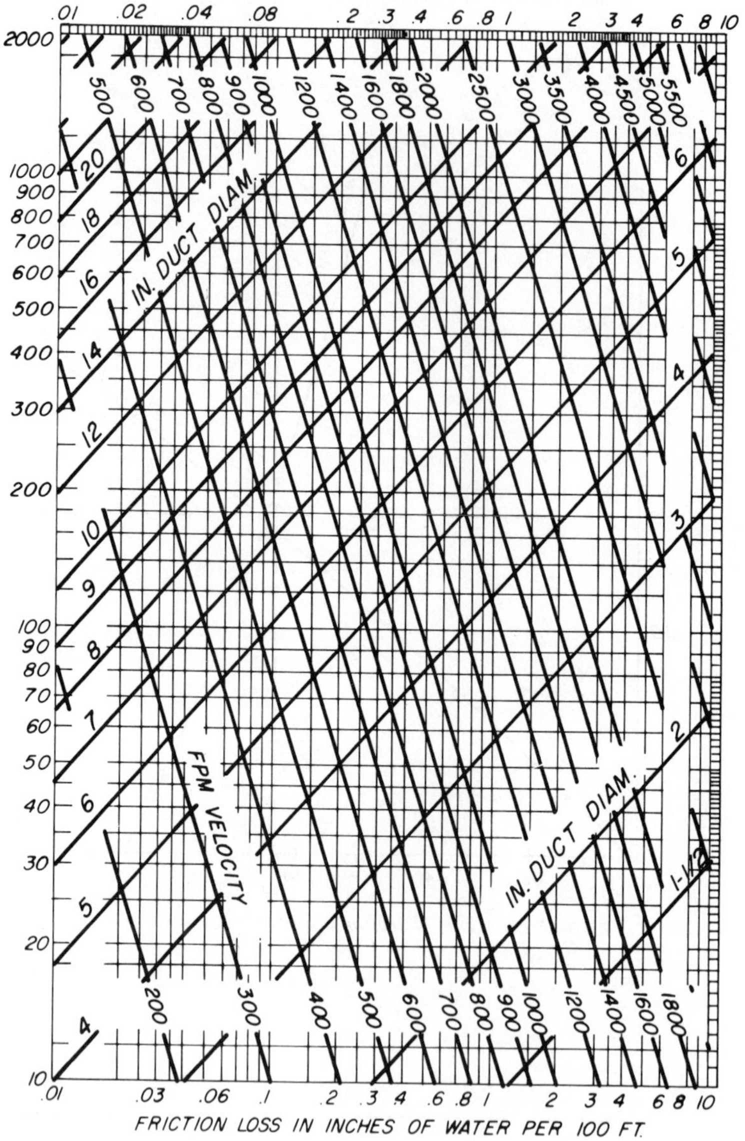

.01 .02 .04 .08 .2 .3 .4 .6 .8 1 2 3 4 6 8 10

500 600 700 800 900 1000 1200 1400 1600 1800 2000 2500 3000 3500 4000 4500 5000 5500

20 18 16 14 12 10 9 8 7 6 5 4 IN. DUCT DIAM.

6 5 4 3 2 1-1/2 IN. DUCT DIAM.

200 300 400 500 600 700 800 900 1000 1200 1400 1600 1800 FPM VELOCITY

CU FT OF AIR PER MINUTE

2000 1000 900 800 700 600 500 400 300 200 100 90 80 70 60 50 40 30 20 10

.01 .03 .06 .1 .2 .3 .4 .6 .8 1 2 3 4 6 8 10

FRICTION LOSS IN INCHES OF WATER PER 100 FT.

Table C-1. Comparative Analysis of Outlets (*Courtesy Carrier Corporation*)

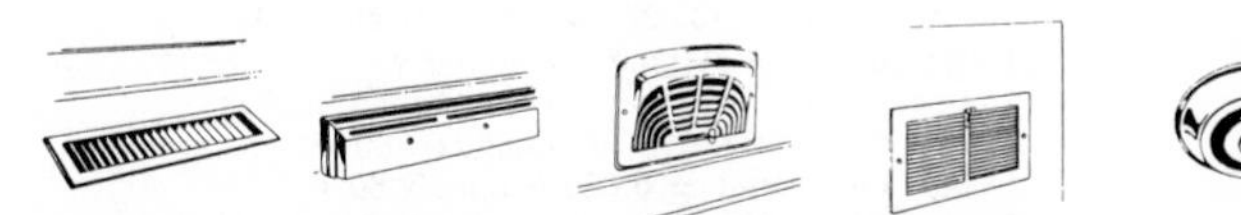

	FLOOR	BASEBOARD	LOW SIDEWALL	HIGH SIDEWALL	CEILING
COOLING PERFORMANCE	Excellent	Excellent if used with perimeter systems	Excellent if designed to discharge upward	Good	Good
HEATING PERFORMANCE	Excellent	Excellent if used with perimeter systems	Excellent if used with perimeter systems	Fair—should not be used to heat slab houses in Northern climates	Good—should not be used to heat slab houses in Northern climates
INTERFERENCE WITH DECOR	Easily concealed because it fits flush with the floor and can be painted to match	Not quite so easy to conceal because it projects from the baseboard	Hard to conceal because it is usually in a flat wall	Impossible to conceal because it is above furniture and in a flat wall	Impossible to conceal but special decorative types are available
INTERFERENCE WITH FURNITURE PLACEMENT	No interference—located at outside wall under a window	No interference—located at outside wall under a window	Can interfere because air discharge is not vertical	No interference	No interference
INTERFERENCE WITH FULL-LENGTH DRAPES	No interference—located 6 or 7 inches from the wall	When drapes are closed, they will clover the outlet	When located under a window, drapes will close over it	No interference	No interference
INTERFERENCE WITH WALL-TO-WALL CARPETING	Carpeting must be cut	Carpeting must be notched	No interference	No interference	No interference
OUTER COST	Low	Medium	Low to medium, depending on the type selected	Low	Low to high—wide variety of types are available
INSTALLATION COST	Low because the sill need not be cut	Low when fed from below—sill need not be cut	Medium—requires wall stack and cutting of plates	Low on furred ceiling system: high when using under-floor system	High because attic ducts require insulation

Conversion Factors

(°C. x 9/5) + 32 = °F.
(°F — 32) x 5/9 = °C.
Liter x 1.05671 = U.S. quarts
Quarts x .946333 = liters
Liters x 61.025 = cubic inches
Gallons x 231 = cubic inches
Kilograms x 2.2046 = pounds
Pounds x 453.59 = grams
Ounces (avdp) x 28.35 = grams
Kilowatts x 1.341 = horsepower
Horsepower x 746 = watts
1 atmosphere = 33.899 feet of water at 39 1°F
1 atmosphere = 760 mm. of mercury
1 atmosphere = 14.7 pounds per square inch
1 cubic foot water = 62.37 pounds @ 60°F
1 cubic inch water = 0.036 pounds@ 60°F
Cubic meters x 35.314 = cubic feet
Cubic feet x 0.02832 = cubic meters
Centistokes x density = centipoises
Pounds/gallon at 20°C. = specific gravity at 20/20°C. x 8.3216
1 centimeter = 0.3937 inches
1 inch = 2.540 centimeter

Cubic Measure

1,728 cubic inches	1 cubic foot
128 cubic feet	1 cord wood
27 cubic feet	1 cubic yard
40 cubic feet	1 ton shipping
2,150.42 cubic inches	1 standard bushel
268.8 cubic inches	1 standard gallon dry
231 cubic inches	1 standard gallon liquid
1 cubic foot	About 4/5 of a bushel
1 Perch	A mass 16½ feet long, 1 foot high and 1½ feet wide, containing 24-2/3 cubic feet.

Miscellaneous

3 inches	1 palm
4 inches	1 hand
6 inches	1 span
18 inches	1 cubit
21.8 inches	Bible cubit
2½ feet	1 military pace

Volume Conversion Tables for English to Metric Systems

Cubic Meters Cubic Yards Cubic Feet	Cubic Feet To:		Cubic Yards To:		Cubic Meters To:	
	Cubic Yards	Cubic Me*ers	Cubic Feet	Cubic Meters	Cubic Feet	Cubic Yards
1	0.037	0.028	27.0	0.76	35.3	1.31
2	0.074	0.057	54.0	1.53	70.6	2.62
3	0.111	0.085	81.0	2.29	105.9	3.92
4	0.148	0.113	108.0	3.06	141.3	5.23
5	0.185	0.142	135.0	3.82	176.6	6.54
6	0.212	0.170	162.0	4.59	211.9	7.85
7	0.259	0.198	189.0	5.35	247.2	9.16
8	0.296	0.227	216.0	6.12	282.5	10.46
9	0.333	0.255	243.0	6.88	317.8	11.77
10	0.370	0.283	270.0	7.65	353.1	13.07
20	0.741	0.566	540.0	15.29	706.3	26.16
30	1.111	0.850	810.0	22.94	1059.4	39.24
40	1.481	1.133	1080.0	30.58	1412.6	52.82
50	1.852	1.416	1350.0	38.23	1765.7	65.40
60	2.222	1.700	1620.0	45.87	2118.9	78.48
70	2.592	1.982	1890.0	53.52	2472.0	91.56
80	2.962	2.265	2160.0	61.16	2825.2	104.63
90	3.333	2.548	2430.0	68.81	3178.3	117.71
100	3.703	2.832	2700.0	76.46	3531.4	130.79

Example: 3 cubic yards = 81.0 cubic feet.
Volume: The cubic meter is the only common dimension used for measuring the volume of solids in the metric system.

Fractions of an Inch

Inch	1/16	1/8	3/16	1/4	5/12	3/8	7/16	1/2
Centimeters	0.16	0.32	0.48	0.64	0.79	0.95	1.11	1.27
Inch	9/16	5/8	11/16	3/4	13/16	7/8	15/16	1
Centimeters	1.43	1.59	1.75	1.91	2.06	2.22	2.38	2.54

Units of Centimeters

Centimeters	0.1	0.2	0.3	0.4	0.5	0.6	0.7	0.8	0.9	1.0
Inches	0.04	0.08	0.12	0.16	0.20	0.24	0.28	0.31	0.35	0.39

Weight[1] Conversion Tables for English to Metric Systems

Number	Metric Ton	Short Ton	Kilograms	Pounds	Grams	Ounces
	Short Ton	Metric Ton	Pounds	Kilograms	Ounces	Grams
1	1.10	0.91	2.20	0.46	0.04	28.4
2	2.20	1.81	4.41	0.91	0.07	56.7
3	3.31	2.72	6.61	1.36	0.11	85.0
4	4.41	3.63	8.82	1.81	0.14	113.4
5	5.51	4.54	11.02	2.67	0.18	141.8
6	6.61	5.44	13.23	2.72	0.21	170.1
7	7.72	6.35	15.43	3.18	0.25	198.4
8	8.82	7.26	17.64	3.63	0.28	226.8
9	9.92	8.16	19.84	4.08	0.32	255.2
10	11.02	9.07	22.05	4.54	0.35	283.5
20	22.05	18.14	44.09	9.07	0.71	567.0
30	33.07	27.22	66.14	13.61	1.06	850.5
40	44.09	36.29	88.18	18.14	1.41	1134.0
50	55.12	45.36	110.23	22.68	1.76	1417.5
60	66.14	54.43	132.28	27.22	2.12	1701.0
70	77.16	63.50	154.32	31.75	2.47	1984.5
80	88.18	72.57	176.37	36.29	2.82	2268.0
90	99.21	81.65	188.42	40.82	3.17	2551.5
100	110.20	90.72	220.46	45.36	3.53	2835.0

Example: Convert 28 pounds to kilograms.
28 pounds = 20 pounds + 8 pounds
From the tables: 20 pounds = 9.07 kg and 8 pounds = 3.63 kilograms
Therefore, 28 pounds = 9.07 kg + 3.63 kg = 12.70 kg

[1]The weights used for the English system are avoirdupois (common) weights. The short ton is 2,000 pounds. The metric ton is 1,000 kilograms.

Length Conversion Tables for English to Metric Systems

Inches / Centimeters / Feet / Meters / Yards / Meters / Miles / Kilometers	Kilometers	Miles	Meters	Yards	Meters	Feet	Centimeters	Inches
	Miles	Kilometers	Yards	Meters	Feet	Meters	Inches	Centimeters
1	0.62	1.61	1.09	0.91	3.28	0.30	0.39	2.54
2	1.24	3.22	2.19	1.83	6.56	0.61	0.79	5.08
3	1.86	4.83	3.28	2.74	9.84	0.91	1.18	7.62
4	2.49	6.44	4.37	3.66	13.12	1.22	1.57	10.16
5	3.11	8.05	5.47	4.57	16.40	1.52	1.97	12.79
6	3.73	9.66	6.56	5.49	19.68	1.83	2.36	15.24
7	4.35	11.27	7.66	6.40	22.97	2.13	2.76	17.73
8	4.97	12.87	8.75	7.32	26.25	2.44	3.15	20.32
9	5.59	14.48	9.84	8.23	29.53	2.74	3.54	22.86
10	6.21	16.09	10.94	9.14	32.81	3.05	3.93	25.40
20	12.43	32.19	21.87	18.29	65.62	6.10	7.87	50.80
30	18.64	48.28	32.31	27.43	98.42	9.14	11.81	76.20
40	24.85	64.37	43.74	36.58	131.23	12.19	15.75	101.60
50	31.07	80.47	54.68	45.72	164.04	16.24	19.68	127.00
60	37.28	96.56	65.62	54.86	196.85	18.29	23.62	152.40
70	43.50	112.65	76.55	64.00	229.66	21.34	27.56	177.80
80	49.71	128.75	87.49	73.15	262.47	24.38	31.50	203.20
90	55.92	144.34	98.42	82.80	295.28	27.43	35.43	228.60
100	62.14	160.94	109.36	91.44	328.08	30.48	39.37	254.00

Example: 2 inches = 5.08 cm

Basic Metric Length Relationships

One Unit (Below) Equals	Millimeters	Centimeters	Meters	Kilometers
Millimeter (mm)	1.	0.1	0.001	0.000,001
Centimeter (cm)	10.	1.	0.01	0.000,01
Meters	1,000.	100.	1.	0.001
Kilometer (km)	1,000,000.	100,000.	1,000.	1.

Square Measure

1 square centimeter	0.1550 square inch
1 square decimeter	0.1076 square feet
1 square meter	1.196 square yard
1 acre	3.954 square rods
1 hectare	2.47 acres
1 square kilometer	0.386 square mile
1 square inch	6.452 square centimeters
1 square foot	9.2903 square decimeters
1 square yard	0.8361 square meter
1 square rod	0.259 acre
1 acre	0.4047 hectare
1 square mile	2.59 square kilometers
144 square inches	1 square foot
9 square feet	1 square yard
30¼ square yards	1 square rod
40 square rods	1 rood
4 roods	1 acre
640 acres	1 square mile

Square Tracts of Land

Acres	Length of One Side of Square Tract, L.F.	Area S.F.
1/10	66.0	4,356
1/8	73.8	5,445
1/6	85.2	7,260
1/4	104.4	10,890
1/3	120.5	14,520
1/2	147.6	21,780
3/4	180.8	32,670
1	208.7	43,560
1½	255.6	65,340
2	295.2	87,120
2½	330.0	108,900
3	361.5	130,680
5	466.7	217,800

Linear Conversions

1 centimeter	0.3937 inches
1 inch	2.54 centimeters
1 decimeter	3.937 inches or 0.328 foot
1 foot	3.048 decimeters
1 meter	39.37 inches or 1.0936 yards
1 yard	0.9144 meter
1 dekameter	1.9884 rods
1 rod	0.5029 dekameter
1 kilometer	0.62137 mile
1 mile	1.6093 kilometers

Volume Conversions

1 cubic centimeter	0.061 cubic inch
1 cubic inch	16.39 cubic centimeters
1 cubic decimeter	0.0353 cubic foot
1 cubic foot	28.317 cubic decimeters
1 cubic yard	0.7646 cubic meter
1 stere	0.2759 cord
1 cord	3.624 steres
1 liter	0.908 dry quarts or 1.0567 liquid quarts
1 dry quart	1.101 liters
1 liquid quart	.09463 liter
1 dekaliter	2.6417 gallons or 1.135 pecks
1 gallon	0.3785 dekaliter
1 peck	0.881 dekaliter
1 hektoliter	2.8375 bushels
1 bushel	0.3524 hektoliter

INDEX